INORGANIC CHROMATOGRAPHIC
ANALYSIS

VAN NOSTRAND REINHOLD
SERIES IN ANALYTICAL CHEMISTRY

Edited by

DR. R. A. CHALMERS

DEPARTMENT OF CHEMISTRY, UNIVERSITY OF ABERDEEN

This series is designed as a coverage of reliable analytical information of value to chemists in research, industry and teaching. Each volume is carefully selected and planned as a modern treatment of a topic of importance to analytical chemists today. New volumes will be added from time to time.

Chemistry of Complex Equilibria	M. T. BECK
Solvent Extraction of Metals	A. K. DE, S. M. KHOPKAR and R. A. CHALMERS
Modern Methods in Organic Microanalysis	JEAN P. DIXON
Errors, Measurement and Results in Chemical Analysis	K. ECKSCHLAGER
Laboratory Handbook of Chromatographic Methods	O. MIKEŠ
Flame Photometry Theory	E. PUNGOR
Gas-Liquid Chromatography	D. W. GRANT
Solution Equilibria in Analytical Chemistry	L. ŠŮCHA and S. KOTRLÝ
Thermal Analysis	A. BLAŽEK

Additional titles will be listed and announced as published

INORGANIC CHROMATOGRAPHIC ANALYSIS

JAN MICHAL Rer. Nat. Dr., Ph.D.

Translation Editor:

JULIAN F. TYSON, B.Sc.

IMPERIAL COLLEGE OF SCIENCE AND TECHNOLOGY, UNIVERSITY OF LONDON

VAN NOSTRAND REINHOLD COMPANY
LONDON
NEW YORK   CINCINNATI   TORONTO   MELBOURNE

VAN NOSTRAND REINHOLD COMPANY LTD.

25—28 Buckingham Gate, London S. W. 1

INTERNATIONAL OFFICES

New York   Cincinnati   Toronto   Melbourne

Library of Congress Catalog Card No. 73-6351
ISBN 442 053 401

Printed in Czechoslovakia by SNTL, Prague

PREFACE

Chromatographic methods are nowadays used extensively for the separation of inorganic components in mixtures, allowing not only the detection of ions after their separation, the identification of individual components in complex mixtures of related inorganic compounds, of different oxidation states, and of various types of complex compounds, but also their quantitative determination. This book is devoted largely to partition and adsorption chromatography. Although separation on ion-exchange resins is also an important chromatographic technique, no special attention has been given to it here as there are already several monographs which cover the field. In this book it is mentioned only where necessary for completeness of the general picture and where the technique is similar to the methods described here in detail (papers impregnated with ion-exchangers, thin layers of ion-exchangers, etc.).

In the introductory part, the principles and the technique of chromatographic analysis are reviewed. A more detailed discussion of these matters will be found in more extensive monographs on the applications of chromatographic methods in the separation of organic mixtures. In this monograph the emphasis is on reviewing the possibilities of separating inorganic compounds, the systems of qualitative analysis of cations and anions, and practical examples of the utilization of chromatographic methods in the qualitative analysis of minerals, metals, alloys, waters and other inorganic materials. The book should serve as a guide to the application of chromatographic methods in inorganic analysis and help analysts to find the most suitable conditions for the solution of problems specific to their laboratories. It therefore includes tables of R_F values of inorganic compounds in the most important solvent systems published up to the present time. It is probably appropriate to mention here that the R_F values can be accurately reproduced only under the conditions described in the paper quoted. That is, the tables are not meant to give absolute R_F values as a basis for identification, but they should help the investigator to choose the best solvent mixture, or the type of solvent best suited for the solution of his problem.

The book also reviews the most important original papers published in periodicals up to the end of 1971. The original intention, to supplement each chapter with recommended literature and thus enable the reader to study more thoroughly even those marginal problems which could not be treated here, had to be abandoned because of lack of space. In spite of this, the literature cited is fairly complete and contains the most important papers in all published areas of inorganic chromatography.

Finally, it is my pleasant duty to express my gratitude to Prof. V. Sýkora, Dr. Z. Šulcek and especially Prof. K. Macek for a detailed revision of the manuscript and many valuable criticisms concerning the conception and treatment of certain parts of the manuscript.

The book aims to help those working in analytical laboratories, in the solution of their practical problems and to facilitate the spread of chromatographic methods in inorganic analysis. However, the author will be glad if the monograph also attracts to these chromatographic methods other highly qualified workers, who will not merely apply them in practice, but will also devote their time to developing them further and finding additional areas of application.

The author

CONTENTS

Chapter 1

INTRODUCTION
TO CHROMATOGRAPHIC METHODS

The history of chromatographic methods goes back to the last century. In 1855 Runge published a book, 'Der Bildungstrieb der Stoffe', where he described his experiments carried out with papers impregnated with various reagents. He used impregnated paper, and applied to it drops of solutions of inorganic compounds which would react with the impregnation reagent, giving very attractive 'pictures'. This work did not represent a great scientific contribution, but it did show how paper could be used for elegant analytical tests. A few years later the discoverer of ozone, Schönbein, published his papers, and became regarded as the founder of capillary analysis. Schönbein observed that when paper was dipped into aqueous solutions of various compounds, water separated from the dissolved material and ascended higher. He made use of this phenomenon for some simple separations in aqueous solution. When a mixture of substances was developed with water, the latter ascended highest and was followed by single substances in the order of their relative diffusion power. Schönbein's pupil, Goppelsröder, developed this method and checked the behaviour of mixtures of various organic and inorganic compounds. He obtained good results when separating some organic compounds, but the results with inorganic compounds were less convincing.

There are other papers from this period which would be considered today to belong to the field of chromatography, but their authors were unable either to interpret or to utilize the results obtained. Thus, for example, Liebig observed as early as 1840 that on filtration of a mixture of salts through a layer of soil, potassium salts were separated from sodium salts. Potassium salts were held in the upper part of the layer and sodium salts in the lower part. This important phenomenon, having particularly great importance in agriculture since it prevented the washing of nutrients into the lower layers of the soil where they could no longer be utilized by the plants, remained without application in analysis. In 1900, Day presented his outstanding results at the 1st International Mineral Oil Congress in Paris. By filtering kerosene through a column of fuller's earth he succeeded in separating single hydrocarbon fractions. The same result was achieved simultaneously and independently by Kvitka. Both recognized correctly the underlying principle of adsorption. However, neither of them substantiated the separation process or explained it theoretically. This was done only in the first years of this century by the Russian botanist Tswett. In 1901 he described the adsorption of plant pigments on proteins, and in his subsequent paper in 1903 he elaborated on his method, explained its principles and proposed a name for it. He called it *chromatography* because this new procedure was applied to the separation of plant pigments. Although the proposed name does not comprise the whole of the present field of chromatography, none of the more suitable names proposed later has been generally accepted, and this very widespread method, indispensable for modern research, preserves its original name.

After Tswett's discoveries came a period when his method was applied only very sporadically. The experts did not trust chromatography, had many objections, and simply did not utilize it for their work. This period of stagnation lasted almost thirty years before there was a sudden new revival and development of the method and also a general acknowledgement of the chromatographic separation principle. The signal for it was given by the work of Kuhn, Winterstein and Lederer in 1931, dealing with the separation of carotenoids. The method rapidly found many applications in the laboratories of organic chemists. However, in inorganic analysis it found a broader application only after 1937 when Schwab and co-workers published a paper on the separation of cations and anions on alumina columns. Schwab's school thus established the basis for a further development of chromatography in inorganic chemistry, which was again stimulated by Flood's use of alumina-impregnated filter paper for the separation of inorganic mixtures. In terms of the sequence of the separated ions his results were almost the same as those of Schwab, but the main significance of his work lay in the fact that he proposed a new technique, i.e. carrying out the separation process in a surface or foil-like arrangement. The same technique was also proposed independently by the Japanese authors Gôto and Kakita, who, however, impregnated their papers in a different manner.

The papers of Consden, Gordon, Martin, and Synge published in the forties announced a revolutionary discovery of fundamental importance for the development of chromatographic methods. In a subsequent development of the principles of Craig's method of countercurrent distribution they used a suitable carrier for the fixation of one phase of the solvent system; first silica gel, then starch and eventually cellulose. For work on the microscale they chose filter paper as a carrier for the stationary phase, thus creating a new, very efficient and simple method of separation — *paper partition chromatography*. Paper had many advantages and it enormously simplified the whole chromatographic process, affording the possibility of a very simple and efficient separation procedure. The importance of the method rapidly outgrew the field of microchemical separations. This made possible an enormous spread of chromatographic methods. The discoverers of paper chromatography, Martin and Synge, were justly awarded the Nobel Prize for chemistry in 1952.

The tremendous spread of chromatographic methods strongly influenced the development of organic analysis, to such an extent that in many cases chromatography is indispensable and even now the only means of achieving an effective separation. The credit for the development of chromatographic methods in inorganic analysis goes to M. Lederer. Other pioneers of this method are Alice Lacourt of the University of Brussels, Pollard and his school of the University of Bristol, Burstall, Linstead and Wells, and others who contributed to the practical applications of the method for separations on cellulose columns. Much attention was also given to this method by Japanese workers, among whom Harasava, Nagai and others proposed a series of very efficient separations. Recently separation on papers impregnated with organic or inorganic reagents, and with ion-exchange materials has become more and more widely used. This field has received special attention from the Italian workers in the school of M. Lederer.

The importance of inorganic chromatographic methods is due to several factors. Ions may be detected easily and without interference after their separation, and single components of complex mixtures of related inorganic compounds or elements in different oxidation states, or types of complex compounds may not only be identified, but also determined quantitatively. In the field of qualitative or even semiquantitative analysis of inorganic compounds, minerals or metals, spectroscopic methods are predominant, but inorganic chromatographic analysis comes to the fore when the character of complex mixtures does not permit a direct determination by the usual methods, and where chromatography makes possible an easy separation into easily determined components. It is especially suitable for use in connection with microchemical problems. For the separation of crude minerals, of metals, alloys, and other materials on a larger scale column chromatography is better because it usually enables a more accurate final determination to be made.

Chromatographic methods help in both the solution of analytical problems and the separation of larger amounts of substances. On sheets, columns, or carriers or adsorbents treated in other ways, substances can be separated in a spectrally pure state. Chromatographic methods can also be applied with advantage to the purification of chemicals, on both laboratory and production scales. Ion-exchangers especially have helped to solve many such problems.

Although separation on ion-exchange resins belongs undoubtedly to chromatographic methods, special attention will not be devoted to it in this book because this field is so well known nowadays and is covered by specialized monographs. Paper electrophoresis also represents an independent area and has a somewhat different application and therefore it also will not be included in the list of chromatographic methods.

Chapter 2

PRINCIPLES
OF CHROMATOGRAPHIC METHODS

For the separation of a mixture by chromatographic methods various modifications are now in use, based on three principles. In order to characterize these principles correctly let us consider first the factors on which chromatographic separation depends.

A common feature of all chromatographic methods is that the separation takes place on a boundary between two phases (solid, liquid or gas, in practically every combination). Chromatographic separation in solutions takes place at the interface, between a solid and a liquid, or between two liquids. The separation of gases takes place at the interface of a solid or liquid phase and a gaseous phase. For inorganic analysis solid-liquid and liquid-liquid chromatography are the most important.

A column or a layer of some material usually serves as the solid phase, and depending on its properties the separation takes place according to the following three principles.

1. The material may have a large adsorptive capacity and the separation is based on differing adsorption coefficients of the substances separated (adsorption chromatography).

2. The material may hold a solvent by sorption, i.e. it retains one liquid phase on its surface and the separation takes place on the basis of differing partition coefficients of the separated substances between the stationary phase and the mobile liquid phase (partition chromatography).

3. The material may be capable of exchanging its own ions for ions present in the mobile solvent and the separation is based predominantly on differing affinity of the separated substances for the exchange material (ion-exchange chromatography).

The classification of chromatographic methods into these three groups should not be considered as being very strict. The situation at the interface of a liquid and a solid phase is usually much more complicated and although one of the principles just outlined may predominate, the influence of other factors cannot be excluded. For the sake of simplicity, however, we shall assume that only one factor is of importance in any one type of chromatography.

2.1. Adsorption Chromatography

Adsorption chromatography, introduced by Tsvett [1, 2] for the separation of plant pigments, was the first chromatographic method to be investigated. A mixture of substances in solution is separated by filtration through a column or layer of a suitable adsorptive material (adsorbent) which should have widely differing adsorptive powers for the different components. The adsorption and desorption should take place rapidly and reversibly. Adsorbents for chromatography fall into two classes: polar or hydrophilic adsorbents, i.e. those having a great affinity for water and other solvents containing the hydroxyl group, and non-polar or hydrophobic adsorbents with a very poor affinity for water. The first group includes alumina and silica gel, the two most commonly used in adsorption chromatography, and also calcium carbonate and sulphate, magnesium oxide, barium and magnesium carbonates and others. The commonest representative of the non-polar adsorbents is charcoal. Numerous organic solvents are used in adsorption chromatography, such as light petroleum, cyclohexane, carbon tetrachloride, ether, benzene, toluene, acetone, chloroform, ethyl acetate, propanol, ethanol, methanol, water, pyridine, etc., and all differ in their power of elution. Choice of a suitable adsorbent and solvent enables the required separation to be achieved.

2.1.1. Adsorption Chromatography on Columns

In column adsorption chromatography the solution containing the mixture to be separated is poured on to a column of adsorbent previously washed with pure solvent. Development (irrigation) is carried out either with the same solvent or with a solvent mixture when this ensures a better separation. The substances may be separated but left in the column (Tsvett's method) or washed out of the column one at a time (elution). Other procedures, known as frontal analysis and displacement chromatography, were introduced by Tiselius [3, 4].

Elution Chromatography

In this case the mixture of substances adsorbed at the top of the column is eluted first by the solvent used to dissolve the mixture. As a rule, this is the solvent with the least elution power. If this solvent is unable to separate and/or elute the mixture, more polar solvents are gradually added until the best separation is achieved. The development of the chromatogram is here directly connected with elution, which can be terminated when the first moving zone, i.e. that containing the least-adsorbed component, reaches the bottom of the column. The column of the adsorbent is then expelled from the column and divided mechanically into zones containing the separated pure components which are then recovered from the adsorbent by extraction with a suitable solvent. This is the classical procedure, called Tsvett's chromatography after its discoverer. It is especially suitable for the separation of coloured substances, when the positions of zones are clearly visible, but it can also be used for the separation of colourless substances if some means of detecting them is available.

The removal of the adsorbent from the glass column is inconvenient and it is preferable to continue the elution until all the zones have been washed out of the column (see Fig. 2.1). Fractions of suitable volume are collected

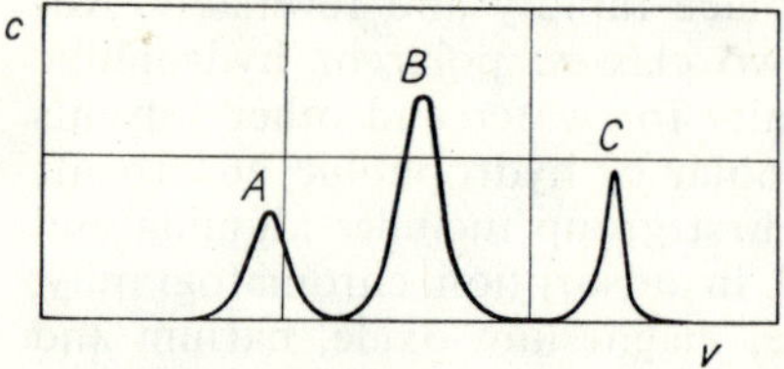

Fig. 2.1. Relation between the concentration of the separated components and the volume of solvent during separation by elution chromatography.

and the separated components are detected or determined in them directly. Manual control of the separation and detection of the components is unnecessary nowadays, as the whole process may be carried out with automatic fraction collection.

Displacement Chromatography

Displacement chromatography is based on the fact that weakly adsorbed components are expelled from the adsorbent by those which are more strongly adsorbed. A mixture of substances is introduced at the top of the column and eluted with a solvent which is adsorbed more strongly than any of the components of the mixture. A displacement of zones thus takes place down the column. The most weakly adsorbed component moves first, followed closely by those more firmly adsorbed, with the solvent itself coming out last. The main difference from elution chromatography is that in the latter the whole column is soaked with solvent and single zones move

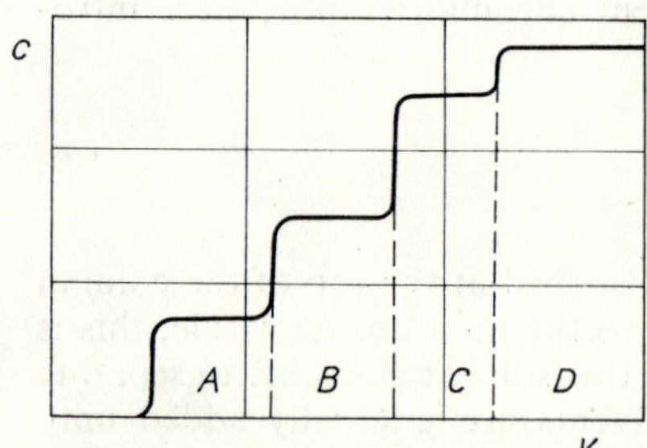

Fig. 2.2. Relation between the concentration of the separated components and the volume of solvent during separation by displacement chromatography.

independently, separated by layers of pure adsorbent if the separation is efficient, whereas in displacement chromatography the solvent moves directly behind the separated zones which also move closely one behind the other and usually overlap slightly.

The dependence of the concentration of the components on the volume of the effluent is shown graphically in Fig. 2.2. Evaluation of this dependence can give the quantitative composition of the mixture. The height of each step is dependent on the adsorption affinity of the corresponding component, and the width is proportional to the quantity of the given substance in the column. However, the example given is an ideal one, and in practice such a case is only attainable with difficulty. Usually neighbouring zones overlap to some extent and their evaluation does not yield the precision demanded by quantitative inorganic analysis.

Frontal Analysis

In frontal analysis the solution to be analysed is poured continuously on to a column of an adsorbent already washed with a solvent, and the eluate is analysed. Pure solvent alone is eluted first and after the so-called retention volume has been run off, the least adsorbed component starts to be eluted. Later the concentration of solute in the eluate increases rapidly at the moment when the second component begins to be eluted. The concentration of the first component in the first 'step' is always higher than in the original solution; when the second component appears in the eluate the concentration of the first falls below the initial concentration. Graphical representation (Fig. 2.3) of the concentration of components versus the volume of the eluate

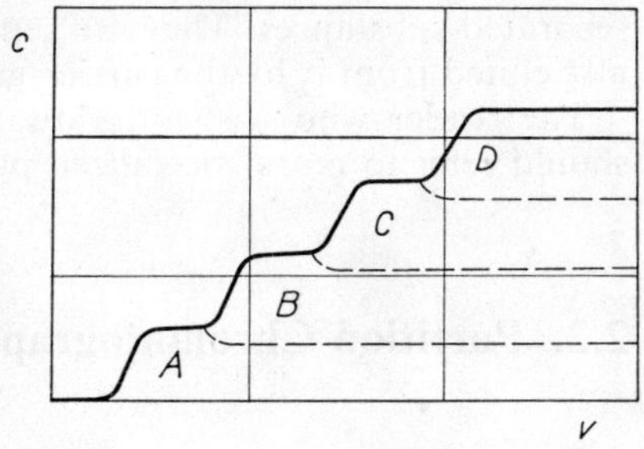

Fig. 2.3. Relation between the concentration of the separated components and the volume of solvent during separation by frontal analysis.

shows this very clearly. Claesson [5] worked out a mathematical expression for the concentration profile of the eluate, according to which it is possible to estimate the quantitative composition of the mixture from the retention volume and the values of the concentration steps.

Frontal analysis, proposed by Tiselius, is suitable for various separations of organic compounds, but with a few exceptions is not generally used in inorganic chromatography.

2.1.2. Adsorption Chromatography on Thin Layers

Thin-layer chromatography has developed tremendously during the last ten years. It is based on the same principle as column chromatography, but the technique is quite different, being closer to that of paper chromatography. The adsorbent is spread as a thin layer on a plate of glass or other material, either as dry powder, thus forming a loose, non-adherent layer, or, more usually, by pouring a suspension of adsorbent over the plate by means of a special applicator, thus giving an adherent thin layer. The suspesion may or may not contain a binder which renders the layer less easily damaged. Adherent layers are more often used because they pemit finer and sharper separation, easier manipulation of the plates, and simpler detection. Another advantage is the possibility of preparing a supply of plates. Various firms offer ready-made plates of very good quality.

2.1.3. Adsorption Gas Chromatography

When the chromatographic separation takes place between a stationary solid phase of great adsorptive power and a mobile gas phase the process is called adsorption gas chromatography. Just as in the case of liquid-liquid adsorption, the techniques of frontal, displacement or elution chromatography can be used. In frontal analysis the differences in flow-rates of the components of a mixture of volatile substances are responsible for the separation. Displacement chromatography is based on the differences in absorptivity of the components and solvent vapours which are introduced into and driven through the column by means of an inert carrier-gas. In the commonest arrangement carrier-gas flows through the column taking with it all separated substances. They are separated on the column and then individually eluted from it by the carrier-gas.

The reader who wishes to study gas chromatography more thoroughly should refer to more specialized publications [6, 7].

2.2. Partition Chromatography

The methods of partition chromatography are very efficient and widely used. They are based on the distribution, or partition, of a solute between two liquid phases. For the sake of simplicity let us assume that the more polar phase is fixed on a solid carrier, and that the less polar phase is percolating through the system. The fixed solvent is called the stationary phase, the moving one the mobile phase. Separation takes place between these two phases and can be described in terms of partition coefficients. In order to explain clearly the process of partition chromatography Craig's model of countercurrent distribution is usually employed because it offers a close analogy to the process and also a graphic representation of the principle of separation.

Let us consider the example of a compound partitioned between equal volumes of two immiscible phases in which its solubilities are in ratio 1 : 4. After one extraction (equilibration by shaking in a separating-funnel) only 80% of the substance is transferred from phase A to phase B. If a quantitative transfer of the compound into phase B is required the extraction of phase B should be repeated several times. After the fourth extraction a

99.84% extraction is attained. Let us imagine now the following experimental arrangement: after extraction, phase A is transferred into another separating-funnel containing solvent B, while pure solvent A is added to the first funnel and the shaking is repeated. The whole procedure is repeated taking care that the portions of phase B remain in the funnels while those of phase A are transferred after each extraction to the next separating funnel. The scheme of this procedure is shown in Fig. 2.4.

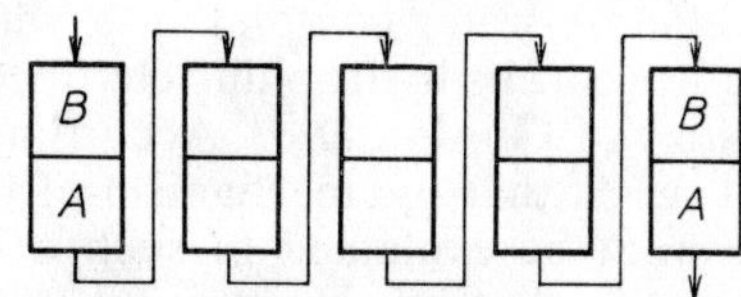

Fig. 2.4. Separation by countercurrent distribution between two immiscible liquids.

Calculation of the true distribution of the compound partitioned between phases A and B in the ratio 1 : 4 in five separating-funnels gives the following values of the sum for the two phases,

1 − 41.0%, *2* − 40.9%, *3* − 15.4%, *4* − 2.54%, *5* − 0.16%

If ten separating-funnels are used the distribution is:

1 − 13.4%, *2* − 30.2%, *3* − 30.2%, *4* − 17.7%, *5* − 6.55%,

6 − 1.58%, *7* − 0.24%, *8* − 0.025%, *9* − 0.008%, *10* − 0.001%

If a second substance partitioned in a 4 : 1 ratio between A and B is similarly treated, the distribution is exactly reversed.

The distribution of the substance between the funnels is best represented graphically, and it is at once obvious that the efficiency of transfer of the solute is dependent on its partition coefficient between the two phases (Fig. 2.5).

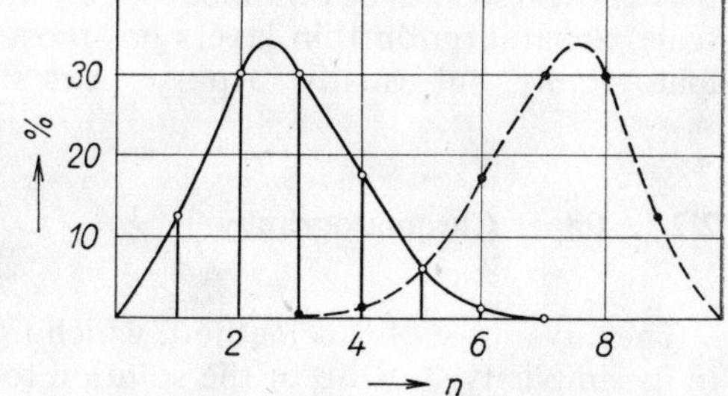

Fig. 2.5. Graphical representation of the distribution of a substance soluble in two immiscible phases in the ratios (-) 1 : 4 and (- -) 4 : 1.

If we now imagine that the more polar phase is fixed on a suitable carrier and that the less polar phase is moving, we can see that a great number of separations may take place, similar to those in the separating-funnels. The efficiency is therefore far greater and the procedure simpler.

Silica gel was the first material used as carrier for the stationary phase. Later cellulose powder was tried, and eventually paper. Gordon, Martin and

Synge [8], who proposed the use of filter paper as the stationary phase carrier in the separation of amino-acids by partition chromatography, assumed that during a chromatographic run water from the mixture of organic solvent and water is retained on the cellulose, while the organic solvent represents the mobile phase. The mechanism of the separation in partition chromatography was later the subject of a number of papers. Hanes and Isherwood [9] in their study of the separation of phosphate esters interpreted the mechanism of the separation as the formation of a cellulose-water complex into which the solute molecule enters as a consequence of its hydrophilic character, and they thus corroborated the assumptions of Gordon *et al.* According to some other papers [10 – 12], however, it seems that the mechanism of the separation in partition chromatography cannot be explained in such a simple manner. The study of quantitative relationships on the boundary between the phases has shown that a layer is formed composed of all components of the solvent system, and that not only partition, but also adsorption and ion-exchange, are involved in the separation process. Nevertheless, for an explanation of the separation process in partition chromatography the original idea of the fixed and moving phases is very convenient.

In practice the selection of a suitable solvent system is of prime importance as this determines the efficiency of the required separation. The choice of the carrier is also important, but it has only a secondary influence on the separation. After the separation the positions of the zones of the separated substances may be detected with good sensitivity by reaction with suitable reagents such as are used for spot tests. Physical methods which make possible the detection of spots on chromatograms without changing the chemical state of the separated substances, such as those based on high-frequency or conductivity measurement, or making use of radioactive labelling, are also in widespread use.

For inorganic partition chromatography the three most important methods are: separation on columns, on paper and on thin-layers. The separation on columns permits work with a larger amount of material and thus generally also a greater accuracy. The separations on paper are simpler and easier; they allow results to be obtained quickly and are also successful on the microscale. Separation on thin layers has most of the advantages of paper chromatography, but usually surpasses it in sharpness and speed of separation.

2.2.1. Paper Chromatography

The advantage of this method, which usually uses filter paper, lies mainly in its simplicity. A drop of the solution to be analysed is applied to a paper strip or sheet which is then developed in a chamber (tank) with a suitable solvent mixture. A graduated cylinder, or even a test tube, may serve as a simple development vessel. When the solvent front has travelled the required distance, the development is interrupted, the front is marked, the chromatogram dried, and the spots are detected by their reactions to chemical tests. The position of a spot, i.e. the ratio of the distances travelled from the starting point by the substance and the solvent front, is a constant under given experimental conditions for each component, and is denoted by R_F, the retardation factor (Fig. 2.6.):

$$R_F = \frac{a}{b}.\tag{2.1}$$

This ratio would be strictly constant only if the spot of the separated substance were a point. As this is not so, R_F is usually expressed as a_2/b,

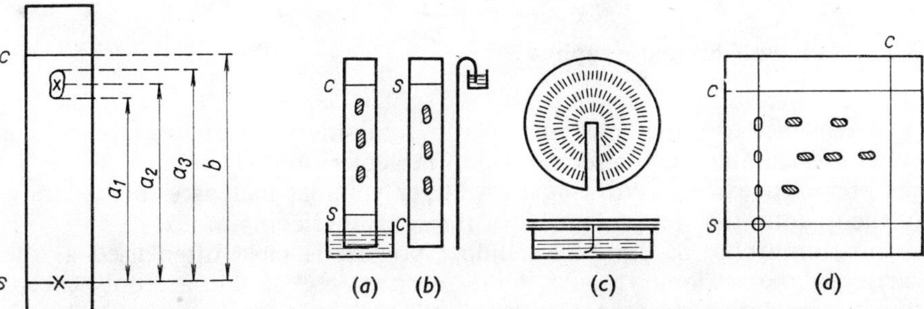

Fig. 2.6. Calculation of the R_F value of a spot.

Fig. 2.7. Various arrangements of paper chromatography. (a) ascending, (b) descending, (c) circular or radial, (d) two-dimensional.

where a_2 is the distance travelled by the middle of the spot, although it would be more suitable to indicate two R_F values for each spot, i.e. the R_F value of its front edge (a_3/b) and that of its terminal edge (a_1/b). From these two values the character of the spot is immediately apparent. A small difference in the R_F values indicates a sharp separation, a large value shows that the spot is diffuse and hence unsuitable for quantitative separation.

The greatest advantage of paper chromatography lies in the possibility of carrying it out in ascending, descending or horizontal arrangements, or even in a consecutive two-dimensional arrangement, facilitating a much better separation of a complex mixture. Ascending, descending and horizontal chromatography differ in the direction of the solvent flow through the paper. In the first the paper imbibes the solvent which moves by capillary ascent (see Fig. 2.7a), in the second the solvent is imbibed by the bent upper part of the paper and moves downwards (see Fig. 2.7b), and in the third the solvent moves in a horizontally placed paper and is either imbibed by capillarity or introduced into the centre of a circular chromatogram (see Fig. 2.7c). Whereas the R_F values in ascending and descending chromatography do not differ significantly and are influenced only by the higher or lower rate of movement of the solvent front caused by gravitation, some authors have shown that in circular chromatography the R_F values are different from those for the other two arrangements. In a two-dimensional arrangement a large square sheet of paper is run first in one solvent in one direction and after drying, in a second solvent in the other (perpendicular) direction. This two-dimensional separation is advantageous because the separation capacities of both solvent systems complement each other, making possible the separation of those mixtures which, in a one-dimensional arrangement, could be separated only with great difficulty or not at all. The principle of two-dimensional separation is shown in Fig. 2.7, *d*.

It sometimes happens that certain ions always have R_F values that are low and very close whatever solvent is used. If such ions are present along with ions of higher R_F value, the technique of 'over-running' can be used. The chromatogram is run until solvent reaches the end of the paper, is dried, and then run again in the same way, and the operation repeated until separation is achieved. The technique is equivalent to the use of a very long paper.

2.2.2. Column Chromatography

In contrast to paper chromatography, partition chromatography on a column is suitable for the separation of larger amounts of substances, up to the preparative scale. Working with larger volumes increases the accuracy of the quantitative determination of the separated compounds.

In column chromatography cellulose powder is most often used as the carrier of the stationary phase. Other carriers, e.g. starch, are only exceptionally employed in inorganic chromatography. The column is filled with the dry material, or with the carrier slurried in the solvent system or its more polar phase. In one series of publications the introduction of the carrier into the column with a fixed aqueous phase is also described. However, on

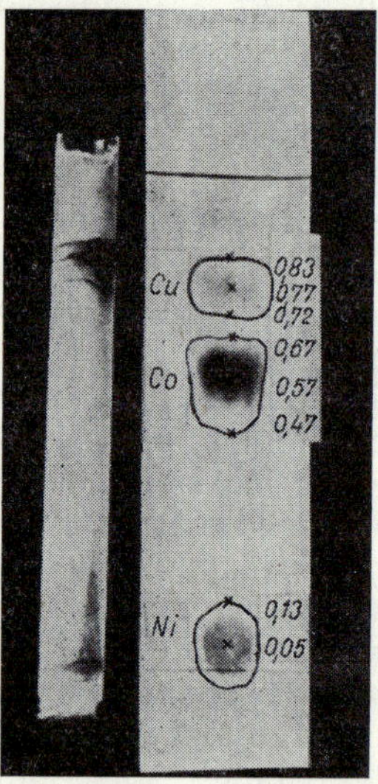

Fig. 2.8. Separation of a mixture of Ni, Co and Cu using acetone —HCl—water solvent Column and paper chromatograms.

closer study of such a procedure it was observed that the solvent system expels water during its passage through the column and that an equilibrium is then attained in the interfacial layer, affected predominantly by the composition of the solvent system.

The most important difference from paper chromatography is the introduction into the column of a much larger amount of solvent. While on paper the solvent is only sucked in by capillary action, the carrier in a column is flooded with the solvent, which affects the chromatographic behaviour of the solutes. If the separation on a cellulose column is carried out by allowing the solvent to ascend into the column by capillary action from below, the results obtained are really comparable with the separation obtained on paper. In Fig. 2.8 the separation of nickel, cobalt and copper in an acetone-hydrochloric acid-water mixture (87 : 8 : 5) on Whatman No. 1 paper is compared with that obtained with a column of dry cellulose. Although some shift in R_F values is observed, especially for cobalt, the character of the separation remains similar. When the separation is carried out on a column in the classical descending manner, however, separations which may be sufficiently effective on paper cannot be well reproduced. Nevertheless, there is a relationship between the behaviour of separated substances on paper and on a cellulose column, which certain authors have tried to express mathematically. Decker and Sano [13] proposed expressions which are valid under certain circumstances and which give the magnitude of the retention volume as a function of R_F (2.2), and also show the relationship between the R_F values on a column and on paper (2.3):

$$V_R = G\left[\gamma_p\left(\frac{1}{R_{F_p}} - 1\right) + \gamma_c\right] \tag{2.2}$$

$$R_{F_c} = \frac{G\gamma_c}{V_R} = \frac{1}{\dfrac{\gamma_p}{\gamma_c}\left(\dfrac{1}{R_{F_p}} - 1\right) + 1} \tag{2.3}$$

where V_R = retention volume, G = weight of cellulose in the column in grams, R_{F_p} = R_F value measured on paper, γ_p = the amount of solvent per g of paper, γ_c = the amount of the solvent per g of cellulose, and R_{F_c} = R_F value measured on the column.

As the R_F values on the column are generally much higher than those on paper, a good separation requires that the compounds to be separated on the column have R_F values between 0.2 and 0.5 on paper. Another necessary condition is that the solvent should flow through the column sufficiently slowly for the two phases always to be in equilibrium.

2.2.3. Thin-Layer Chromatography

In thin-layer partition chromatography the stationary-phase carrier is usually silica gel or powdered cellulose. The chromatograms are developed in the same way as paper chromatograms, and the ascending arrangement is used almost exlusively. Its main advantages are speed and sharpness of separation.

2.2.4. Gas Partition Chromatography

Although gas chromatography is not the subject of this publication, a short mention of partition gas-chromatography should be made here. We describe as partition gas-chromatography separations in a gas-liquid system in which the liquid phase is fixed on a suitable carrier which itself has very weak adsorptive properties, and where the gas phase is mobile, flowing over the fixed liquid. The technique used in this method is elution chromatography.

2.3. Chromatography on Ion-Exchange Material

Chromatography on ion-exchangers represents the third basic group of chromatographic methods. Here the coulombic electrostatic attractive forces are made use of for separations. Generally, the function of an ion-exchanger can be illustrated by the following example: the ion B^+ is added to a column of ion-exchanger RA and is retained by it. However, as electroneutrality must be preserved, ion A^+ is liberated from the exchanger. Hence, the separation takes place according to the scheme

$$RA + B^+ \rightleftharpoons RB + A^+ \tag{2.4}$$

If the exchanger captures cations from the solution it is called a cation-exchanger, if anions are exchanged, it is an anion-exchanger.

The exchange of ions takes place on their diffusion from the solution to the ion-exchanger particles where they are captured, simultaneously releasing an equivalent amount of other ions. Equilibrium is attained rapidly between the solid and the liquid phases, characterized by Eqs. (2.5) and (2.6):

$$A_1^+ + B_2^+ \rightleftharpoons A_2^+ + B_1^+ \tag{2.5}$$

$$K = \frac{[A_2^+]\,[B_1^+]}{[A_1^+]\,[B_2^+]} \tag{2.6}$$

Subscript 1 indicates the ions in the exchanger, and subscript 2 the ions in the solution. The condition necessary for a successful application of a given exchanger is that equilibrium be attained rapidly.

In practice many different ion-exchangers are used. Natural inorganic substances, usually aluminosilicates, were used first. They served as models for the preparation of synthetic inorganic compounds having similar structures to the natural compounds but a greater capacity [14, 15]. However, organic ion-exchangers represent a much more important group. Among natural materials some types of sulphonated coal have been found useful in certain instances, but exchangers of the synthetic resin type find very wide application. Polycondensate and polymeric exchangers are presently produced by many famous companies and all monographs on ion-exchangers and their applications list the material and their characteristics.

Just as in adsorption and partition chromatography, so also in ion-exchange chromatography both columns and sheets can be used. Columns are usually used when the required ion is simply captured from the sample solution and subsequently eluted from the column and determined. At present this is the most important application of ion-exchangers. Both common types of exchangers and most types of liquid exchangers may be used in the form of thin layers or impregnated papers. Commercial impregnated papers can be used with advantage.

As several monographs have been devoted to ion-exchangers in inorganic analysis, treatment of ion-exchange column chromatography will be omitted in this book. However, a brief mention will be made at the appropriate places of the separations on ion-exchange papers or on thin layers of exchangers.

REFERENCES

1. TSVETT, M. S. *Ber. Deutsch. Botan. Ges.*, **24,** 234, 316, 384 (1906).
2. TSVETT, M. S. *Biochem. Z.* **58,** 225 (1914).
3. TISELIUS, A. *Arkiv. Kemi Mineral. Geol. B,* **14,** 22 (1940).
4. TISELIUS, A. *Advan. Colloid Sci.* **1,** 81 (1942).
5. CLAESSON, S. *Disc. Faraday Soc.* **7,** 34 (1949).
6. PURNELL, J. H. *Gas Chromatography* Wiley, New York (1962).
7. KAISER, R. *Gas-Chromatographie* Geest and Portig, Leipzig (1960).
8. GORDON, A. H., MARTIN, A. J. P. and SYNGE, R. L. M. *Biochem. J.* **37,** xiii (1943).
9. HANES, C. S. and ISHERWOOD, F. A. *Nature* **164,** 1107 (1949).
10. MICHAL, J. and ACKERMANN, G. *Talanta* **11,** 441 (1964).
11. MICHAL, J. and ACKERMANN, G. *Talanta* **11,** 451 (1964).
12. MICHAL, J. and ACKERMANN, G. *Talanta* **12,** 171 (1965).
13. DECKER, P. and SANO, I. *Z. Physiol. Chem.* **300,** 252 (1955).
14. VESELÝ, V. and PEKÁREK, V. *Talanta* **19,** 219 (1972).
15. PEKÁREK, V. and VESELÝ, V. *Talanta* **19,** 1245 (1972).

Monographs

BLASIUS, E. *Chromatographische Methoden in der analytischen und präparativen anorganischen Chemie (unter besonderer Berücksichtigung der Ionenaustauscher)* Enke Verlag, Stuttgart (1958).
HAIS, I. M. and MACEK, K. (editors) *Paper Chromatography* Publishing House of the Czechoslovak Academy of Sciences, Prague (1963).
LÁBLER, L. and SCHWARZ, V. (editors) *Chromatografie na tenké vrstvě* Nakladatelství Českoslov. Akad. Věd, Prague (1965).
LEDERER, M. *Progrès récents de la chromatographie II. Chimie minérale* Hermann, Paris (1952).
MIKEŠ, O. *Laboratory Handbook of Chromatographic Methods,* Van Nostrand, London (1966).
POLLARD, F. H. and McOMIE, J. F. W. *Chromatographic Methods of Inorganic Analysis,* Butterworth, London (1953).
RANDERATH, K. *Dünnschicht-Chromatographie* Verlag Chemie, Weinheim (1965).
SAMUELSON, O. *Ion Exchange Separations in Analytical Chemistry* Wiley, New York (1963).
STAHL, E. (editor) *Dünnschicht-Chromatographie, ein Laboratoriumshandbuch* Springer Verlag, Berlin (1967).
ŠINGLIAR, M. *Plynová chromatografie* SVCTL, Bratislava (1961).

Chapter 3

TECHNIQUES
OF CHROMATOGRAPHIC ANALYSIS

The character of the solid phase (adsorbent, carrier, ion exchanger) is of prime importance and therefore great attention should be devoted to the choice of this material. The group of solvents employed in adsorption and partition chromatography is practically the same, but the criteria according to which single solvents are used differ. It is necessary, therefore, to discuss briefly the choice of solvents.

3.1. Adsorption Chromatography

3.1.1. Adsorbents

The commonest adsorbent in both inorganic and organic chromatography is alumina. Its evident advantage lies in its capacity to bind cations as well as anions. However, when it is used for the separation of inorganic ions pure adsorption does not play an exclusive role, as formation of insoluble compounds may also take place on the surface and the mechanism of separation is much more complicated. Ion-exchange is also involved, i.e. the exchange of H^+ or Na^+ ions for heavy metal ions, and of the OH^- ion for anions. To simplify the mechanism of separation of inorganic ions on alumina it is advantageous to use the complexes of the ions with organic reagents. In this manner the adsorptive capacity of alumina for organic molecules is utilized and the same adsorbents, solvents, and solvent systems can be used as in organic chromatography.

When inorganic ions are separated on alumina, single zones are not as a rule separated by layers of 'empty' adsorbent material, but are connected with each other. Only when an ion is weakly adsorbed and moves rapidly through the column does a separation take place in which the zones are completely separated. It is, therefore, often advantageous to use weak adsorbents.

Brockmann and Schodder [1] proposed procedures for testing the adsorption activity of alumina, and for preparing alumina of various degrees of activity. Their test is based on experimental determination of the sorption capacity of alumina for a series of azo dyes, for example azobenzene, *p*-methoxyazobenzene, Sudan Yellow, Sudan Red, *p*-aminoazobenzene and *p*-hydroxyazobenzene. A benzene solution of dyes of neighbouring adsorptive properties, diluted with light petroleum, is filtered through a standard size column of the alumina to be tested, and it is observed whether each dye remains on the upper or the lower part of the column, or passes into the filtrate. The positions of the various dyes indicate the activity of the alumina. The most active alumina is prepared by heating small amounts of technical alumina in an iron crucible over the direct flame of a gas burner. The activity is then decreased, if necessary, by moistening, best done by exposure to atmospheric humidity. As the alumina absorbs water its activity slowly decreases. Alumina of various degrees of activity, according to Brockmann, contains the following amounts of water: I — 0%, II — 3%, III — 6%, IV — 10%, and V — 15%.

Among adsorbents used in inorganic chromatography, zinc sulphide, calcium sulphate and silica gel should be mentioned. The last two have uses similar to those of alumina, but zinc sulphide operates by causing precipitation of heavy metal sulphides from solution, because they are less soluble than ZnS itself. Gold, and the platinum metals with which such an effect cannot be expected, are also separated, and precipitation takes place even in the presence of complexing reagents, such as 8-hydroxyquinoline and violuric acid.

In the case of polar adsorbents, the most common of which is alumina, adsorption depends on the character of the adsorbed substance, primarily on the number of polar groups in the molecule, its polarizability, the number of double bonds, the dipole moment of the molecule, etc. This is, however, only the case when the inorganic ion is bound as a complex with an organic reagent.

On non-polar adsorbents the effects mentioned play only a minor role and adsorption depends predominantly on the size of the molecule, a factor which can be made use of only for the separation of ions complexed with larger organic molecules. Practically the only representative of this type of adsorbent is active charcoal on which complexed ions may be separated from free ions. The sorption should take place from polar solvents and desorption should be carried out with non-polar solvents.

3.1.2. Solvents

Water is usually used for the separation of ions on alumina. Schwab and co-workers [3, 4] investigated the sorption of a series of ions from neutral aqueous solutions and established the sequence of adsorbability. However, if use is made of organic chelating agents, the solvent becomes an important factor in the separation process. When the solvent passes through the adsorbent it also is adsorbed and the separation taking place is a result of the ratio of the strengths of sorption of the separated substances and of the solvent. Trappe [5—7], after a thorough investigation of solvent-alumina interactions, was able to classify solvents in a series according to their

Table 3.1. Eluotropic Series According to W. Trappe

Solvent	Solubility of water in the solvent (%)	Dielectric constant
Water	unlimited	81.1
Methanol	unlimited	31.2
Ethanol	unlimited	25.8
2-Propanol	unlimited	26.0
1-Propanol	unlimited	22.2
1-Butanol	7.9	19.2
Ethyl acetate	8.6	6.1
Ether	7.5	4.4
Chloroform	1.0	5.1
Benzene	0.08	2.24
Toluene	0.05	2.3
Carbon tetrachloride	0.08	2.25

decreasing power of adsorption on hydrophilic adsorbents, a sequence now called the eluotropic series and which gives the order in which solvents should be used for the elution of a chromatographic column (see Table 3.1). The upper members of this series represent the strongest eluents for alumina. In the table are listed only those solvents most frequently used for the separations of complex compounds by adsorption chromatography. From the table it can also be seen that the adsorption strength of the solvents increases as their solubility in water and their dielectric constants diminish. For separations on charcoal the order of solvents is almost entirely reversed.

The solvents mentioned can be applied either alone or in combination. The solvents should be applied sequentially from the least to the most polar, for a polar adsorbent. As they also are bound on the adsorbent, it would be useless to reverse the order, because after the column has been eluted with a more polar solvent, a less polar solvent would be without effect. In the case of non-polar solvents the opposite is true.

Mention should also be made of the purity of the solvents used, especially their water-content. Water in non-polar solvents increases their polarity and the result of the separation no longer corresponds to the suppositions made. Commercial ether usually contains a certain amount of ethanol, which increases its polarity. Therefore great attention must be paid to the purity of solvents used, because an impurity may appreciably affect the result of a separation. Purification procedures are described in monographs on laboratory techniques.

3.1.3. Equipment

The basic equipment for adsorption chromatography is a column, usually a simple glass tube of suitable size. If Tsvett's method is to be used, the tube is open at the lower end and the packing is easily accessible; for elution chromatography the column ends with a stop-cock. Some basic types of chromatographic columns are shown in Fig. 3.1. In (*a*) a glass column [8] is fitted with a stopper at the top, through which the stem of a separating funnel passes, and has a cotton-wool plug at the bottom. The bottom of the column is pressed on a Witt plate in a glass funnel connected to a Büchner flask and a pump. The chromatographic tube is taken out of the funnel and the column

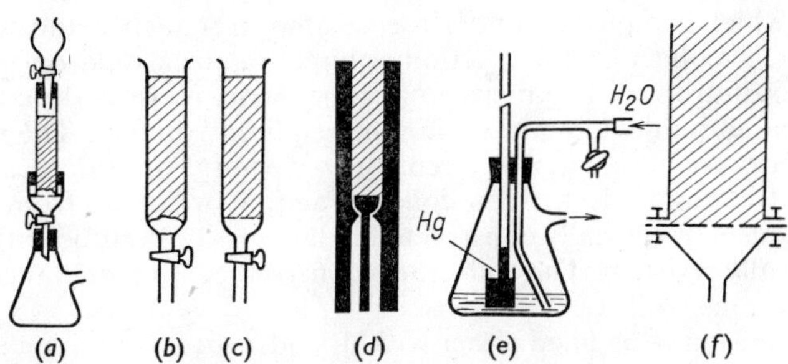

Fig. 3.1. Fundamental types of chromatographic columns.
(*a*) column according to Woelm, (*b, c*) columns for the elution technique, (*d*) column for a micropreparative separation, (*e*) device for creating pressure in a chromatographic column, (*f*) chromatographic kit (sectional column) according to Winterstein.

of adsorbent pushed out of it. For elution chromatography the columns represented in Fig. 3.1, (*b* and *c*) are suitable. These are tubes narrowed at their bottom end and closed either with a cotton-wool plug (*b*) or a Witt plate (*c*). For micropreparative work the arrangement shown in Fig. 3.1*d* is convenient. This is especially suitable for the classical procedure and makes it very easy to push out the adsorbent from the column. The application of reduced pressure to accelerate the chromatography is sometimes dangerous, especially if volatile solvents are used. When the suction is too strong the lower part of the adsorbent column becomes dry and sometimes breaks. If necessary the elution can be speeded up by application of slight pressure at the top of the column (Fig. 3.1*e*). A suitable device, easily put together, is recommended by Herout [9]. If larger amounts of solvent are required (as in preparative work) composite columns are used. Winterstein's design is shown in Fig. 3.1*f*.

The ratio of length to diameter for a column of adsorbent may vary between 5 : 1 and 15 : 1, but is usually around 10 : 1. The larger the diameter of the column, the narrower are the zones, but at the same time the solvent flow through the column becomes more irregular, and irregular zones may be formed.

The substances separated on the column usually have to be made visible. In classical chromatography the extruded column of adsorbent is either sprayed with the detection reagent or the reagent is applied with a brush, usually as a streak. In elution chromatography the identification of the separated zones is usually much simpler. Single eluate fractions may be analysed for the component sought, or a continuous automatic recording device may be employed. These devices are usually based on some physico-chemical method, for example photometry, high-frequency conductimetry, or polarography. Lovett and Wright [10] described a device for the continuous measurement of the UV-absorption of an eluate, based on a double-beam system comparing the eluate and pure solvent. A similar arrangement was also described by McClendon [11]. More commonly the eluate is collected in a large number of tubes (fractions) in which the components are then determined. Various fraction collectors have been constructed [46]. They are usually composed of a rotating disc or drum provided with a large number of vessels which are pushed under the column in sequence and for a preset time. In older collectors the fraction volume was measured on the basis of the number of drops falling in a known time. More recently the construction has been based on devices measuring the required volumes. The most accurate collectors are based on measurement by weight. For example, Shandon's automatic fraction collector can collect fractions weighing from 1 to 60 g. The mechanism of specially constructed balances is both sufficiently sensitive and chemically resistant. The collector is constructed for simultaneous collection of fractions from two columns.

The columns may be filled either with dry adsorbent, or with a suspension of adsorbent in some suitable solvent. It is also possible to pour dry adsorbent into a column already filled with the solvent. The first method is handicapped by the difficulty of achieving a regular distribution of the dry adsorbent in the tube: an irregular column causes irregular separations. It is necessary to assist regular sedimentation of adsorbent in the tube by tapping the tube gently from both sides durng the filling, or by connecting the tube to a vibrator. In the second method (filling the tube with a suspension) separation of larger particles from finer ones often takes place in the column and the zones are irregular. Therefore the third method of filling is considered best i.e. pouring the dry adsorbent in a thin stream into the column filled with the solvent, provided that the solvent is run out regularly and slowly during the operation. If the column is filled with dry adsorbent, it must be washed briefly with the solvent before application of the material for separation. The latter must be put on the top of the wet adsorbent column. In all instances the mixture to be separated should be introduced onto the column as a concentrated solution. Only after this concentrated solution has been sucked into the adsorbent column is a small amount of solvent introduced. If necessary the same solvent is also used for washing the walls of the tube. When this solvent has also been adsorbed a larger amount of solvent is added carefully, without disturbing the top surface of the column.

In classical chromatography the column of adsorbent should be expelled from the column immediately after development. If the column begins to dry out, it is usually very difficult, if not impossible, to extrude the column undamaged. Split columns (cut in two halves longitudinally and held together with adhesive tape) are very convenient for this procedure.

3.2. Partition Chromatography

3.2.1. Materials

As carriers for the stationary phase, silica gel or cellulose are most often used for columns, and various sorts of paper in paper chromatography.

Silica gel for chromatography is commercially available but it can be prepared easily in the laboratory from water-glass. Pitra and Štěrba [12—14] investigated silica gel, especially from the point of view of its use in adsorption chromatography, and proposed methods for its preparation and evaluation. They recommend the fourfold dilution of a commercial water-glass with water to give a solution of sp. gr. 1.070, addition of 30 ml of conc. ammonia per litre, followed by filtration through a column of cation-exchanger (in H^+-form) to eliminate the cations present. The sol of silicic acid flowing from the column should have pH 2.5—3.0. As soon as the pH begins to increase above 3.0 the filtration should be stopped immediately, because it means that the resin is exhausted and requires regeneration. If prepared for partition chromatography where maximum pore size is required (for a maximum sorption of water as stationary phase), the pure sol is poured rapidly and with thorough stirring into a 10% ammonium bicarbonate solution. The gel formed is filtered off under suction, washed with water and dried at a temperature at which the added ammonium bicarbonate decomposes. For fractionating dried and ground silica gel the authors proposed a method based on differential sedimentation.

The most important carrier in partition chromatography is cellulose, the material of plant cell membranes, consisting of polymeric D-glucose.

Depending on the method of preparation, these chains contain 50—500 glucose units and one glucuronic acid unit:

Cellulose chains are oriented longitudinally and form submicroscopic crystalline micelles containing a network cross-linked by hydrogen bonds between primary and secondary alcohol groups.

On exposure to water vapour, cellulose takes up 20—25% of water. However, the weight increase is not the same for organic solvents. In Fig. 3.2

the saturation of Whatman cellulose by the vapour of water, methanol, ethanol, 1-propanol, 2-propanol, 1-butanol and 2-butanol is shown. It can be seen that the weight increase diminishes with increasing molecular weight of the alcohols. The alcohols used were distilled several times and had a very low water content (0.5−1%). It is evident from studies of the composition of the solvent bound on the surface of the cellulose that both water and solvents are absorbed, and that the concentration ratio of water and alcohol

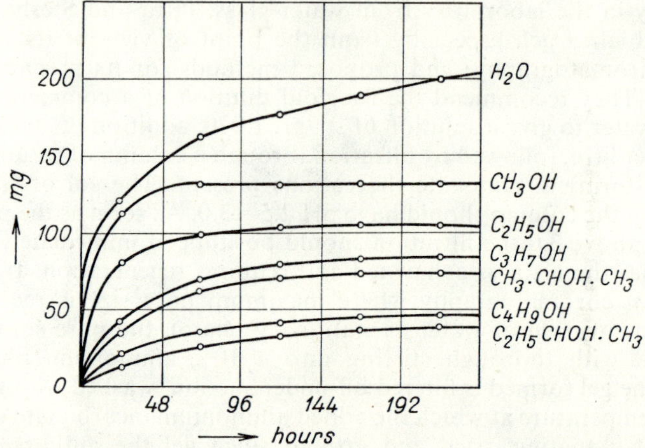

Fig. 3.2. Relation between the weight increase in Whatman cellulose and time during saturation with water, methanol, ethanol, 1-propanol, 2-propanol, 1-butanol and 2-butanol vapour.

bound on the cellulose is quite different from that of the original solvent. The water content in the bound liquid increases with alcohol chain-length until in the case of butanol it increases from the original 1.16% to 90−98% (Table 3.2). Michal and Ackermann [15−17] studied in detail the behaviour of cellulose in solvent vapours, and also the situation at the interface of the

Table 3.2. Sorption of Alcohol Vapours by Cellulose

Solvent	Weight increase g	Water g	Alcohol %
Water	0.1733	0.1732	—
Methanol	0.1227	0.0106	91.4
Ethanol	0.1030	0.0296	71.3
1-Propanol	0.0687	0.0221	67.8
2-Propanol	0.0670	0.0278	58.5
2-Butanol	0.0428	0.0343	18.4
1-Butanol	0.0311	0.0305	2.0

solid and liquid phases during the chromatographic process, and in contrast to a number of authors they concluded that the situation cannot be explained simply. An interstitial layer is formed on the boundary of both phases, composed of all components of the solvent system. It is the change in concentration of single components of the solvent system in the interstitial

Table 3.3. Types of Chromatographic Papers

Type*	Number	Weight g/m²	Thickness mm	Flow rate	Remarks
W	1	90	0.17	medium	
W	2	100	0.18	slow	
W	3	185	0.37	medium	
W	3MM	180	0.32	medium	
W	4	95	0.19	fast	
W	17	100	1.0	medium	
W	20	95	0.17	very slow	
SS	2040a	85	0.18	fast	similar to W 4
SS	2040b	120	0.23	fast	similar to W 4
SS	2043a	85	0.18	medium	similar to W 1
SS	2043b	120	0.23	medium	similar to W 1
SS	2045a	90	0.16	very slow	
SS	2045b	120	0.18	very slow	
SS	2071	650	0.75	very slow	
SS	2727	700	1.15	fast	
SS	2181	2500	4.0	medium	
FN	1	90		fast	similar to W 4 and SS 2040a
FN	2	120		fast	similar to W 4 and SS 2040b
FN	3	90		medium	similar to W 1 and SS 2043a
FN	4	120		medium	similar to W 1 and SS 2043b
FN	5	90		very slow	similar to SS 2045a
FN	6	120		very slow	similar to SS 2045b
FN	7	150		fast	
FN	8	280		fast	similar to SS 2247

*W = Whatman (Reeve Angel, London).
SS = Schleicher and Schüll (Dassel, German Federal Republic)
FN = Niederschlag (VEB Specialpapierfabrik, German Democratic Republic).

layer in relation to the composition of the solvent system itself which influences the chromatographic separation.

For column chromatography Whatman Ashless Cellulose Powder Standard Grade (Reeve Angel and Co) or Schleicher and Schüll Zellulosepulver No. 123, 123a (especially the purified 123), and 124 (with negligible adsorptive power) are most often used. Among chromatographic papers the products of Reeve Angel and Co. (Whatman), Schleicher and Schüll, and also Spezialpapierfabrik, VEB Niederschlag, are best known. A review of these papers is given in Table 3.3. Among other well-known kinds of paper those of Munktell, of Macherey, Nagel and Co., and also of Binzer should be mentioned.

3.2.2. Solvents

The choice of a suitable solvent system in inorganic partition chromatography is rather difficult. Unfortunately, there are no general rules, and experience is more valuable and effective than logical or theoretical considerations. That little progress has been made in this direction may be attributed to the complexity of the separation process. If it could be supposed that only the more polar components were bound by the carrier, the separation processes could be studied much more easily. In practice, we must suppose that an interstitial layer is formed on the boundary of the solid and the liquid phase, and that the composition of this layer depends primarily on the composition of the solvent system used. Even in the case of two-component systems the situation is still rather complicated. If the composition of this boundary layer in the system methanol-water in various ratios is investigated [18], it is found that the water content in the interstitial layer increases in proportion to the water content in the solvent system, but that this layer does not have the same composition at various places along the paper, even when experimental conditions exclude any evaporation or condensation during the chromatographic process. As solvent flows through the carrier the water content of the latter reaches that of the original solvent system relatively slowly. Figure 3.3 shows the dependence of the water content in the interstitial layer at a certain point on the chromatographic paper strip on the water content in the solvent system methanol-water. The curves for various development times show that equilibrium is attained after approximately 72 hours, the difference in composition of the layer and the solvent then being minimal for all mixtures of alcohol and water. From this it follows that in practice chromatographic analysis is done under non-equilibrium conditions and that the composition of the interstitial layer is not constant but changes with the amount of solvent passed, though very slowly. It is also quite evident from Fig. 3.3 that during the chromatographic process the composition of the interstitial layer differs relatively little from the composition of the solvent system if the water content of the latter is greater than about 50%. It is also of interest that the composition of the interstitial layer is virtually unaffected by variations in temperature.

Although our theoretical understanding of the separation process is at present rather limited, several facts are known from experience. Thus, for example, Pollard and McOmie [19] enumerate ten factors affecting inorganic chromatographic analysis:

1 — the nature of the inorganic compounds which are to be chromatographed; their nature can be changed by the presence of some component of the solvent system (formation of complexes),
2 — the nature of the polar solvent (usually) which can be partly or totally miscible with water,
3 — the water content of the solvents and the humidity of the laboratory,
4 — the acidity of the mobile phase and of the original solution of the salts,
5 — the disturbing effect of one ionic species on the mobility and partition of another,
6 — the presence of other ions,
7 — the duration of chromatography,
8 — the temperature,
9 — the attainment of equilibrium in the presence of complexing agents in the solvent system,
10 — the properties and the state of the cellulose carrier.

Among these factors the influence of the complexing effect of the solvent system and the acidity of the mobile phase play especially important roles. A suitable choice of pH for the solvent system can create optimum conditions for the fomation of complex ions and thus often facilitate satisfactory separation.

The solvents used in partition chromatography do not differ substantially from those in adsorption chromatography. In addition to a series of alcohols,

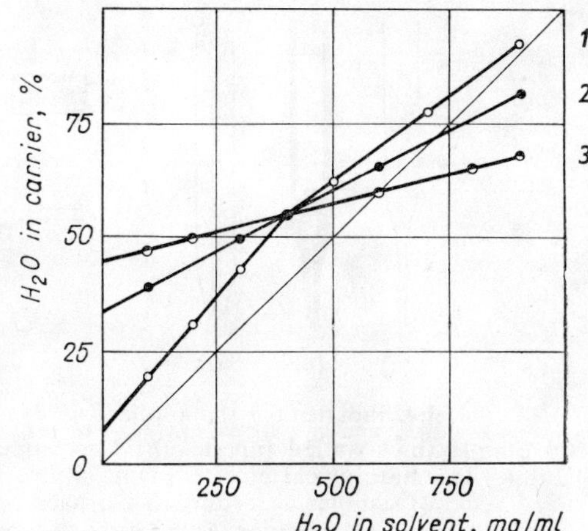

Fig. 3.3. Relation between the water content in a given zone of the chromatogram and the water content in the solvent system during ascending chromatography with methanol — water. Time of development: *1* — 30 minutes; *2* — 24 hours; *3* — 48 hours.

ketones such as acetone, ethyl methyl ketone, acetylacetone, etc., ethers such as diethyl and methyl ethyl or dioxan, esters such as methyl and ethyl acetate, and chloroform, carbon tetrachloride and other solvents are used. The factors affecting the chromatographic process and influencing the choice of the solvent system will not be further discussed here. The main reason for this is the prevailing empiricism in the choice of experimental conditions.

However, in order to orient the reader rapidly in the possibilities offered by paper partition chromatography in inorganic analysis a detailed review of procedures for the separation of single ions is given, and the R_F values of all elements in most of the various solvent systems described in the literature are tabulated.

3.2.3. Apparatus

For column partition chromatography on cellulose or silica gel, the apparatus is usually similar to that for elution adsorption chromatography. However, the filling of the column with cellulose powder should be done more carefully than in the case of inorganic adsorbents. The column is best filled with dry cellulose with constant tapping or vibration of the glass tube. After the addition of each portion of cellulose the layer in the tube should be pressed gently with a piston (a round metallic plate, perforated if possible, fixed on a rod). Before chromatography the column is washed with the most polar component of the solvent system used, and then with the least polar component. Alternatively the column may be filled with a suspension of cellulose in the solvent mixture to be used later for the separation.

The techniques, used in paper partition chromatography are usually simpler. The arrangements used depend on the direction of solvent flow. If the solvent moves vertically downwards, the method is called descendent, if upwards it is called ascendent. If the solvent moves in a horizontal plane, the method is called horizontal chromatography, the most common modifi-

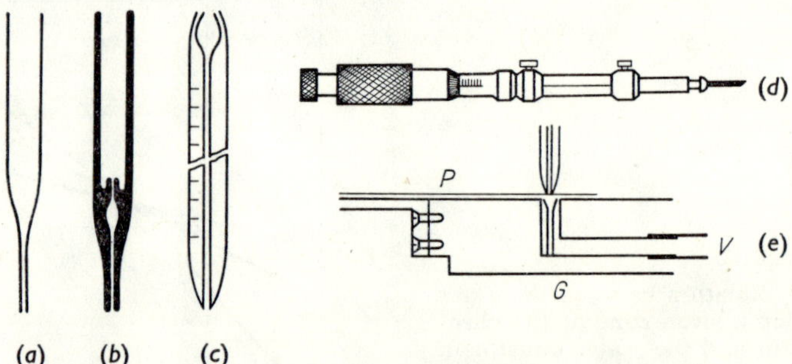

Fig. 3.4. Pipettes for the application of mixtures onto chromatograms.
(a) Simple thick-walled pipette. (b) Micropipette according to Meinhard and Hall [20]. (c) Haematological pipette. (d) Agla pipette, (e) Arrangement for the application of samples according to Gorbach and Gallob-Hausmann [27].
(P = paper; G = Gorbach-type heater; V = pump).

cation being circular chromatography. Ascendent and descendent chromatography can be carried out either in one direction only (one-dimensional chromatography) or in two perpendicular directions (two-dimensional chromatography). The arrangement for chromatography of larger amounts of substances, aimed at the isolation of single components, is rather different and is called preparative paper chromatography.

For ascending and descending chromatography the shape of the paper used depends primarily on the size and shape of the chromatographic tank. Narrow strips, broad bands or complete sheets may be used. The place of the application of the sample to be separated, called the start, is marked beforehand with a soft pencil, usually 2 – 3 cm from the edge for the ascending system and 6 – 10 cm from the edge for descending chromatography. The sample is applied on the start with a suitable pipette. For qualitative

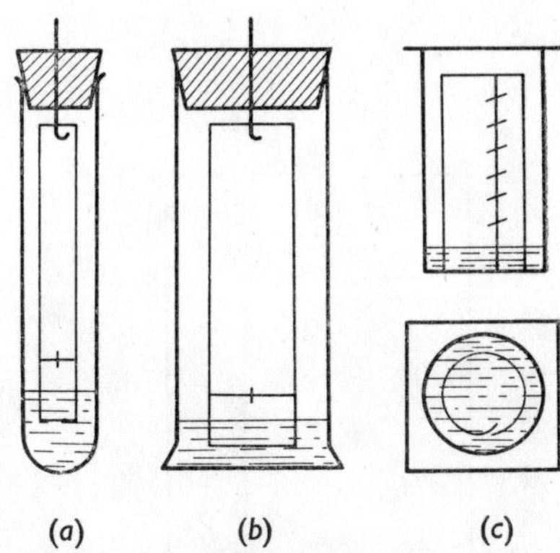

Fig. 3.5. Some arrangements for ascending chromatography.
(*a*) Separation in a test-tube. (*b*) Separation on a strip of paper in a graduated cylinder. (*c*) Separation on a sheet of paper in a chromatographic chamber (tank).

or semiquantitative separations a strong glass capillary is employed (Fig. 3.4*a*), or a micropipette according to Meinhard and Hall [20] (Fig. 3.4*b*). Haematological pipettes (Fig. 3.4*c*) are also useful. Pipettes can easily be calibrated by weighing the liquid released from the capillary. For precise quantitative work micropipettes, microburettes, or micrometer syringes, all of which can be used to deliver an accurately measured volume, are necessary. The best known type is the 'Agla' (Fig. 3.4*d*). Camag market a set of micropipettes with calibrated tips for 2.5 – 10 µl volumes, and a holder for convenient handling. After application of the sample to the start, the sample solvent must be evaporated. A hair-dryer or an infrared lamp is usually used.

A series of papers has been devoted to the problem of mechanization or automation of the application of samples to the paper [21 – 28]. As the sample should be applied at the start in the form of a very small spot, devices have been constructed for simultaneous addition and evaporation of the sample. In Fig. 3.5 an arrangement according to Gorbach and Gallob-Hausmann [27] is shown schematically with which 100 µl of sample may be applied as a spot of 1 – 2 mm diameter, over a period of 15 – 30 minutes. This method of application prevents the formation of a ring of material on the circumference of the applied spot, which usually happens with repetitive addition. A consequence of the ring-form of spot is unsatisfactory separation, especially if the spots are initially rather large, and the R_F values similar.

The ascending technique is usually used for preliminary tests. The equipment for this consists of a stoppered test-tube [28], the stopper of which carries a narrow paper strip with its lower edge immersed in the solvent (Fig. 3.5*a*). Alternatively, paper strips 2 cm broad are used in 250-ml graduated cylinders serving as chromatographic tanks (Fig. 3.5*b*). A glass rod is inserted through the stopper closing the cylinder and bent at its lower end into a hook on which the paper is hung. The lower edge of the paper is immersed in the solvent by pushing the glass rod downwards.

For ascending chromatography on paper sheets (usually when routine work or two dimensional chromatographs is being done) the arrangement shown in Fig. 3.5c has been found suitable. The paper sheet is rolled into a cylinder with the two edges sewn or clipped together, and introduced into a cylindrical tank containing a small volume of solvent mixture [29]. If clips

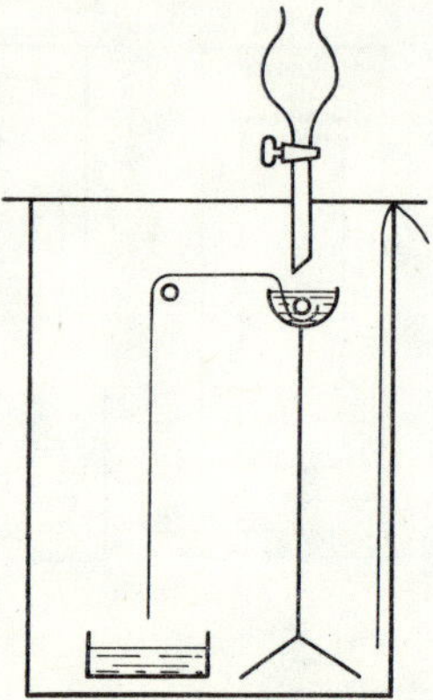

Fig. 3.6. Equipment for descending chromatography.

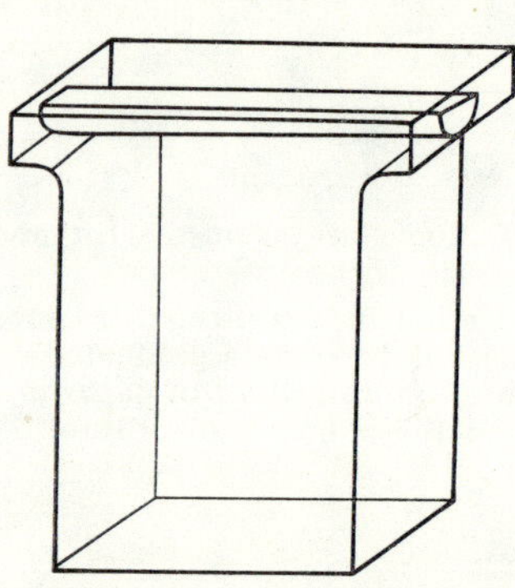

Fig. 3.7. Chromatographic chamber according to Macek and Hais for descendingchromatography.

are used they must be made of plastic, because the acidic solutions that are often used in inorganic chromatography would dissolve some iron, etc. out of metal clips.

The scheme for descending paper chromatography is represented in Fig. 3.6. A strip or sheet of chromatographic paper is fixed in a trough (usually of glass) by means of a heavy glass rod. A suitable amount of solvent is poured into the trough. Usually it is first necessary to saturate the chromatographic chamber with the vapours of the solvent system, and only then to start the chromatography by introducing the solvent into the trough. The saturation is often done by putting Petri dishes or other vessels in the bottom of the tank and filling them with solvent, but it is more efficient if the walls of the tank are lined with filter paper wetted with the solvent. A suitable chromatographic chamber designed by Macek and Hais, enabling troughs to be inserted easily, is shown in Fig. 3.7.

Of the many different arrangements for circular paper chromatography, two are to be preferred. In one, a drop of sample is applied directly to the centre of the paper disc and dried. A narrow strip is then cut in the direction of the centre of the disc and bent downwards so that the paper has the shape of a drawing pin [30, 31] (Fig. 3.8). The narrow strip is pushed through a hole in a glass cover into an underlying Petri dish filled with the solvent, and the paper is then covered with a second glass plate. The arrangement according to Zimmermann and Nehring [32] may be found more convenient

(Fig. 3.8*b*). The paper is fixed between the cover and the lower part of a desiccator and the solvent is introduced from above through the cover, from a separating-funnel. A paper wick is inserted into the centre of the paper in order to drain off the excess of solvent. This arrangement permits presaturation of the paper with the solvent vapours (this presaturation is called equilibration).

Although for preparative work column chromatography is more suitable, it is sometimes convenient to separate large amounts of substances on paper, since no problems arise from preparation of the column. Two arrangements

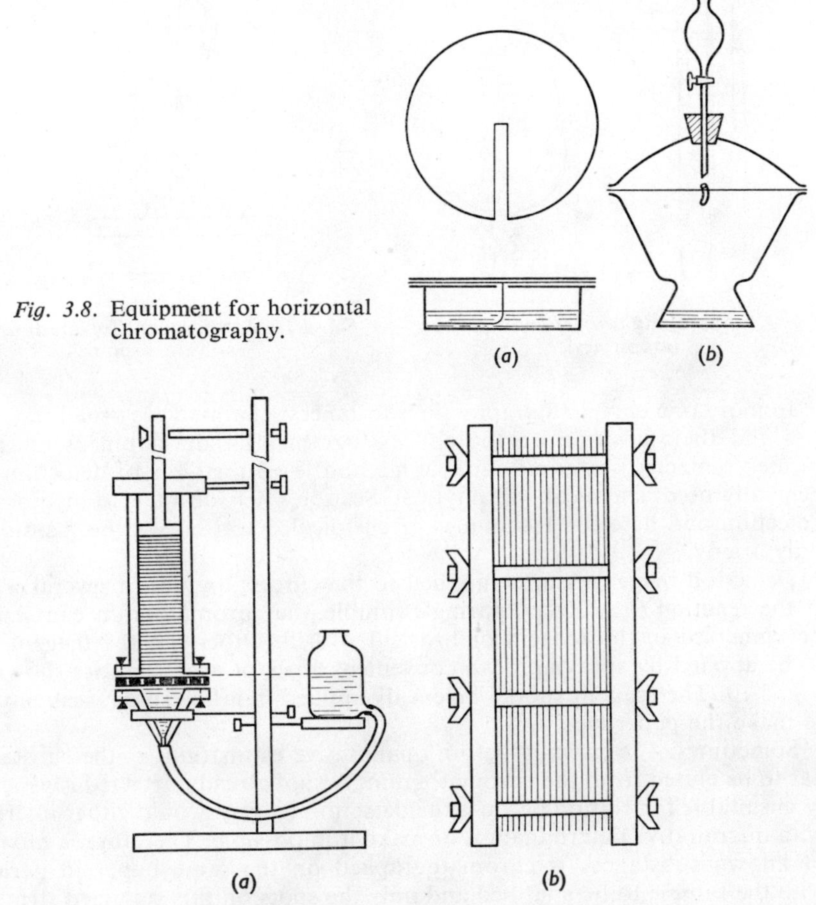

Fig. 3.8. Equipment for horizontal chromatography.

(a) (b)

(a) (b)

Fig. 3.9. Equipment for preparative paper chromatography.
(*a*) arrangement according to Roth ('chromatopile')
(*b*) Porter's 'chromatopack'.

are shown in Fig. 3.9, those according to Roth [33] ('chromatopile', Fig. 3.9*a*) and to Porter [34] ('chromatopack', Fig. 3.9*b*). In Roth's arrangement discs of filter paper are firmly pressed together to form a pile (column), covered with a glass cylinder, and used for ascending chromatography.

The solvent in the reservoir should be level with the lowest disc of the pile. In Porter's arrangement a pack of chromatographic papers is pressed between two stainless steel plates. Ascending development is again used. The simplest method of preparative paper chromatography is separation on chromatographic cardboard, i.e. especially thick filter paper, having a relatively high capacity. The best known are Whatman No. 17 and Schleicher and Schüll No. 2181 papers.

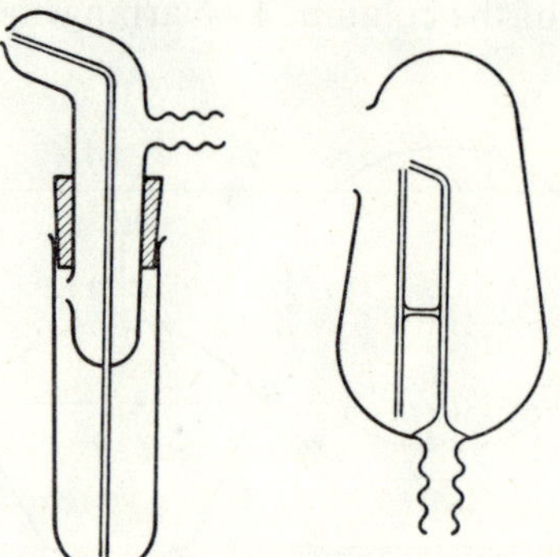

Fig. 3.10. Spraying devices for detection (atomizers).

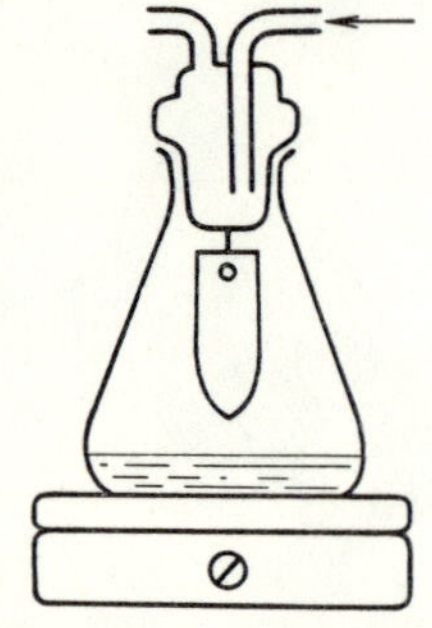

Fig. 3.11. Arrangement for elution with solvent vapour.

In inorganic chromatography the substances separated are usually colourless and their position on the chromatogram after development must be located (detected) by some suitable method. Two methods of detection are generally used: chemical and physical. Section 4.1 is devoted to methods of detection and here the techniques of chemical detection will be mentioned only briefly.

Detection reagents can be applied to the chromatogram in several ways. If the reaction product is sparingly soluble, the chromatogram can simply be immersed in the reagent or brushed with it. Otherwise the reagent has to be applied by spraying. Two convenient types of atomizers are shown in Fig. 3.10. The reagent should be evenly applied in a fine spray, just enough to make the paper wet.

Sometimes — especially before quantitative estimation — the substance has to be eluted from the chromatogram. A spot already detected is generally unsuitable for elution because the detection reagent would either interfere with quantitative determination or make it impossible. Therefore, a mixture of known substances is chromatographed on the same paper in parallel with the sample to be analysed and only the spots on this standard strip are detected. The corresponding zones of the sample chromatogram are then cut out and eluted. In the simplest method the part cut out is boiled with water or acid in a glass beaker. If this method is unsuitable the cut out zone is put into an extractor (Fig. 3.11) and eluted with the vapour of a suitable solvent. If it is necessary to elute with a minimum amount of solvent, the zone is fixed between two glass plates, with an edge of the paper projecting. Immersion of the opposite edge of the plates into a solvent causes elution by capillary ascent into the protruding edge and concentration of the substance in a relatively small area from which it can be eluted with a very small amount of solvent. Several other methods have also been proposed [35 – 45].

3.3. Thin-Layer Chromatography

It has already been mentioned that thin layers may be used in adsorption and partition chromatography, and even for ion-exchange chromatography. The technique is the same for all three.

The materials used in adsorption or partition chromatography have already been discussed, but a few remarks should be added on their use in thin-layer chromatography.

Thin layers are classified as adherent (fixed) and non-adherent (loose). The former are prepared by pouring a suspension of a suitable material on to a plate (usually of glass), and the latter by spreading a dry adsorbent on the plate. For partition chromatography non-adherent layers are made of suitably prepared silica gel of approx. 0.1 mm particle size. For adherent layers thick suspensions of adsorbents or carriers (usually alumina or silica gel) in suitable solvents, usually water, are prepared. The adsorbent in the suspension may be used alone, or accompanied by some binder such as gypsum. For the preparation of such layers very finely ground material is necessary.

Common alumina is more often used for chromatography on non-adherent layers, its grain-size being ~ 0.1 mm; its activity is regulated and determined either by methods common in column chromatography (see p. 16) or according to Heřmánek and co-workers [47]. Commercial preparations include Lachema alumina for chromatography, Merck Aluminiumoxyd, Camag alumina (type DS-O without binder and DS-5 with 5% gypsum as binder), and Woelm alumina, usually without binder [offered in three qualities: acid (pH ~ 4), neutral (pH ~ 7.5), and alkaline (pH ~ 9)].

Silica gel is the commonest material for the preparation of adherent thin layers. Commercial preparations include Spolana (Velvary) silica gel CH, Merck preparations, e.g. Kieselgel H without binder (5 – 25 μ particle size) and Kieselgel G with 13% gypsum as binder (same particle size), Camag silica gel, types D-0 and DS-0 without binder and type D-5 with 5% gypsum, and Woelm silica gel without binder.

The materials for the preparation of thin layers are not always sufficiently pure, often containing inorganic ions such as iron, which are detrimental to separations of inorganic mixtures. They must therefore be purified, which is best done directly by developing the layers with a mixture of methanol and hydrochloric acid (9 : 1), drying and, if necessary, reactivating.

Ready-made thin-layer plates have become very popular because they have a uniform thickness and give more reproducible results. The best known plates are those of Merck and Eastman Kodak, coated with alumina or silica gel for adsorption chromatography, or with cellulose for partition chromatography, the 'Polygram' plates of Macherey-Nagel & Co. with layers of silica gel, cellulose MN 300, MN 300 PEI, alumina or polyamide, and the Schleicher and Schüll 'Selecta' plates with silica gel or cellulose layers. 'Silufol' plates (Kavalier, Votice, Czechoslovakia, distributed also by Serva, Heidelberg, German Federal Republic), consist of cardboard coated on both sides with aluminium foil on which a thin layer of silica gel with starch as binder is fixed. The presence of starch complicates certain detection methods requiring corrosive reagents or elevated temperatures, but the ease of handling the material and the remarkable cheapness make these plates very attractive for many purposes.

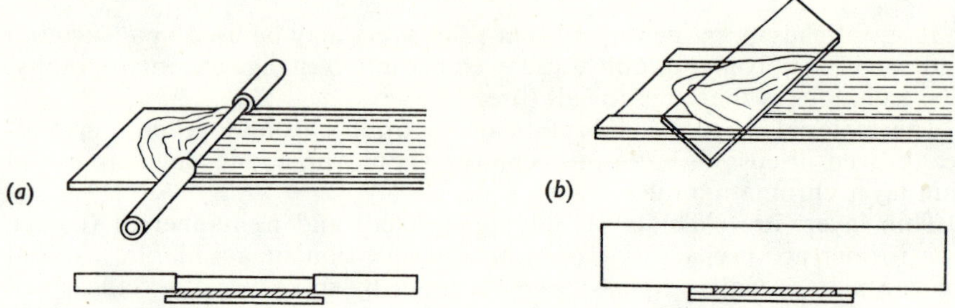

Fig. 3.12. Simple devices for making layers of required thickness.

For the preparation of thin layers in the laboratory either simple laboratory tools or one of the numerous commercial devices, known as applicators, can be used. The non-adherent layers can be prepared most simply by means of a glass rod or tube on which two small pieces of rubber tube (rings) are placed the required distance apart (Fig. 3.12a), or by using a glass plate with raised edges (Fig. 3.12b). When adherent layers are to be prepared these simple devices are not satisfactory and other home-made [48, 49] or commercial applicators must be used. Among the latter Stahl's applicator (supplied by Desaga), which enables the thickness of the layer to be adjusted, is most commonly used. Desaga also sell kits of basic equipment for thin-layer chromatography, containing the applicator, glass plates of various sizes, drying stands, special pipettes, chromatography tanks, and the material for making the thin layers. Camag also make excellent equipment.

The prepared thin-layer plates must be dried before use, preferably in a drying oven with controlled temperature and a forced draught, in which rapid drying may be achieved even at room temperature. Such an apparatus is described by Barnard and Turner [50].

For the application of samples the same pipettes and devices can be used as in paper chromatography, but more care must be taken with non-adherent layers, which may be damaged by rough handling and the quality of separation impaired. The development must also be done very carefully if damage to the layer, or even its falling off the plate, are to be avoided. For this reason the use of chromatographic tanks of suitable size is recommended, so that the plate need not be inclined at more than 20°.

Adherent layers can also be developed in a vertical position. Various types of chromatographic chambers are commercially available, but those used for paper chromatography are often satisfactory. In thin-layer chromatography saturation of the atmosphere in the chromatographic tank is very important. Tanks with a very small capacity have been developed, saturation of which is very simple and rapid. They are called 'sandwich chambers', because the glass plate with the layer forms the rear wall and the chamber is completed by a second plate, a lid, and a frame determining the width of the chamber. The inner side of the front plate can be coated with a layer which when wetted rapidly saturates the chamber. Such a chamber, produced by Camag, is shown in Fig. 3.13.

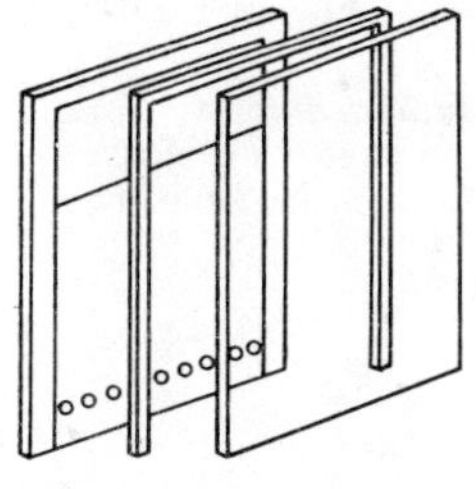
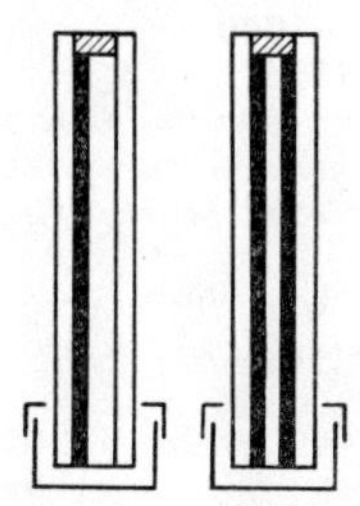
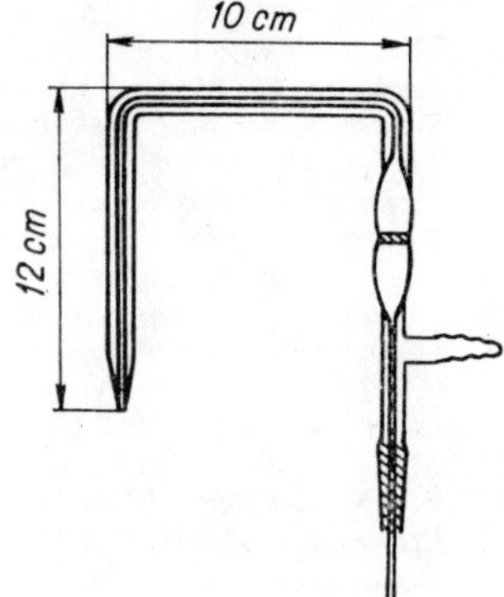

Fig. 3.13. Camag 'sandwich-chamber'.

Fig. 3.14. Extractor of zones according to Matthews and co-workers [51]

The methods of detection on adherent layers are those used in paper chromatography, but as the plates are usually of glass and the layers of inorganic material, much more drastic reagents can often be used. For non-adherent layers the reagents should be applied (sprayed) very carefully in order to prevent damage to the layer. Spraying is best done from some distance, with very fine droplets of the reagent falling on to the layer by their own weight.

Sometimes the separated substance should be taken off the chromatogram after detection. For this purpose a number of devices have been proposed, generally constructed for removal of a zone of layer by suction. Such aspirators either suck the material from the layer directly into a flask, where it is then extracted with a suitable solvent, or retain it on a sintered glass filter from which the substance is subsequently eluted, with a suitable solvent. An illustration of such a device is given in Fig. 3.14.

3.4. Ion-Exchange Materials

The commonest method of separation on ion-exchangers is column chromatography. Usually the ions or group of ions are simply retained on the column, the impurities are washed out, and the required material is then eluted from the column with another solvent.

If a true separation during elution is required, the thin-layer technique may be used, or chromatography on papers impregnated with ion-exchangers. The papers can be purchased or impregnated in the laboratory. Papers impregnated with Dowex and Amberlite resins are available. For the impregnation of papers various substances have been used, among them tri-n-octylamine, di-(2-ethylhexyl)orthophosphoric acid, zirconium phosphate, etc.

REFERENCES

1. BROCKMANN, H. and SCHODDER, H. *Ber.* **74**, 73 (1941).
2. *Chromatographie unter besonderer Berücksichtigung an Papierchromatographie* Merck, A.G., Darmstadt.
3. SCHWAB, G. M. and JOCKERS, K. *Angew. Chem.* **50**, 546 (1937).
4. SCHWAB, M. and DATTLER, G. *Angew. Chem.* **50**, 691 (1937).
5. TRAPPE, W. *Biochem. Z.* **305**, 150 (1940).
6. TRAPPE, W. *Biochem. Z.* **306**, 316 (1940).
7. TRAPPE, W. *Biochem. Z.* **307**, 97 (1941).
8. WOELM, M. [according to BLASIUS *Chromatographische Methoden in der analytischen u. präparativen anorganischen Chemie* Enke-Verlag, Stuttgart (1958)].
9. HEROUT, V. in *Chromatografie. Sborník prací, Přírodovědecké vydavatelství* Prague (1952).
10. LOVETT, S. and WRIGHT, W. C. *Chem. Ind. London*, 1433 (1961).
11. McCLENDON, J. H. *Anal. Biochem*, **3**, 94 (1962).
12. PITRA, J. *Chem. Listy* **56**, 495 (1962).
13. PITRA, J. and ŠTĚRBA, J. *Chem. Listy* **56**, 544 (1962).
14. PITRA, J. and ŠTĚRBA, J. *Chem. Listy* **57**, 389 (1963).
15. MICHAL, J. and ACKERMANN, G. *Talanta* **12**, 171 (1965).
16. MICHAL, J. and ACKERMANN, G. *Talanta* **11**, 441 (1964).
17. MICHAL, J. and ACKERMANN, G. *Talanta* **11**, 451 (1964).
18. MICHAL, J. and ACKERMANN, G. *unpublished work*.
19. POLLARD, F. H. and McOMIE, J. F. W. *Chromatographic Methods of Inorganic Analysis*, p. 59. Butterworths, London (1953).
20. MEINHARD, J. E. and HALL, N. F. *Anal. Chem.* **21**, 185 (1949).
21. PORTEOUS, J. W. *J. Chromatog.* **2**, 58 (1959).
22. MATOUŠEK, V. *Chem. Listy* **57**, 270 (1963).
23. ZALISHVILI, M. M. and SHRAIBMAN, F. O. *Biokhimiya* **25**, 570 (1960).
24. MERZ, W. *Mikrochim. Acta* 474 (1957).
25. STOCKER, H. R. *Helv. Chim. Acta* **46**, 2050 (1963).
26. VAN DER SIJDE, D. and DE FLINES J. *J. Chromatog.* **2**, 436 (1959).
27. GORBACH, G. and GALLOB-HAUSMANN, G. *Mikrochim. Acta* 102 (1962).
28. ROCKLAND, L. B. and DUNN, M. S. *Science* **111**, 332 (1950).
29. WILLIAMS, R. H. and KIRBY, H. *Science* **107**, 481 (1948).
30. RUTTER, L. *Nature* **161**, 435 (1948).
31. RUTTER, L. *Analyst* **75**, 37 (1950).
32. ZIMMERMANN, G. and NEHRING, K. *Angew. Chem.* **63**, 556 (1951).
33. ROTH, L. *Mikrochim. Acta* 582 (1959).
34. PORTER, W. L. *Anal. Chem.* **23**, 412 (1951).
35. TRYHORN, F. G. and CURRY, A. S. *Nature* **178**, 1180 (1956).
36. SERZISKO, E. and SCHIERITZ, M. *Chem. Techn.* **8**, 207 (1956).
37. SKOLIK, J. *J. Chromatog.* **3**, 273 (1960).
38. STANFORD, F. G. *J. Chromatog.* **11**, 125 (1963).
39. REITH, W. S. *Nature* **179**, 580 (1957).
40. BROICH, J. R. *J. Chromatog.* **5**, 365 (1961).
41. BROER, Y. *Bull. Soc. Chim. Biol.* **43**, 283 (1961).
42. BOOTH, V. H. *J. Chromatog.* **6**, 95 (1961).
43. CANNY, M. J. *J. Chromatog.* **3**, 496 (1960).
44. LEWIS, B. *Chem. Z.* **128**, 2290 (1957).
45. MIERAU, H. *J. Chem. Z.* **128**, 10571 (1957).
46. VEČEREK, B. *Chem. Listy* **57**, 337 (1963).
47. HEŘMÁNEK, S., SCHWARZ, V. and ČEKAN, Z. *Collection Czech. Chem. Commun.* **26**, 3170 (1961).
48. BARBIER, M., JÄGER, H., TOBIAS, H. and WYSS, E. *Helv. Chim. Acta* **42**, 2440 (1959).
49. MACHATA, D. *Mikrochem. Acta* 79 (1960).
50. BARNARD, W. E. and TURNER, N. A. *J. Chromatog.* **29**, 296 (1967).
51. MATHEWS, J. S., BEREDA, A. L. and AGUILERA, A. P. *J. Chromatog.* **9**, 331 (1962).

Chapter 4

METHODS OF DETECTION
AND DETERMINATION
OF SEPARATED COMPONENTS

Chemical methods of detection involve exposing the chromatograms to a suitable reagent (e.g. by spraying) which forms a coloured compound with the required ion. Physical methods are much more varied. Especially important are those based on the use of radioactive isotopes labelling the separated compounds. Among other methods, colorimetric ones are most often used, but a number of methods based on polarography and spectrography have been described, as well as others based on conductimetric, refractometric, interferometric and high-frequency measurements, etc.

Chromatographic methods are often used for the isolation or separation of compounds before their quantitative determination. After the elution of the separated substances from the paper or column they can be determined in the eluates by any method which is applicable to the quantity of material available. Paper and thin-layer chromatography facilitate the application of physico-chemical methods for the direct determination of separated compounds on chromatograms, without the necessity of eluting them first. For this purpose photometric and spectrophotometric methods are mainly used, but many other forms of measurement have been employed.

4.1. Detection

4.1.1. Chemical Methods

Success in chemical detection depends primarily on the choice of reagent. Specific or highly selective reagents [1] may be chosen, or, on the other hand, reagents giving a clear and sufficiently sensitive colour reaction with several ions. Selective reagents are useful if it is uncertain whether the required element has been completely separated from other elements which would make its detection by a non-selective reagent impossible. Non-selective reagents are preferred when the components have been well separated and a large number of components can be detected with one reagent.

Gánti and Novak [2] proposed a widely applicable method. A dried chromatogram is sprayed with a 0.5% benzidine solution in glacial acetic

Table 4.1. Reactions of Cations with Some General Detection Reagents

Element	Colour								
	1	2	3	4	5	6	7	8	9
Aluminium	+	−	+	+	+	+	fl	−	−
Antimony	−	−	(III) +	+	(III) +	+	−	wV	−
Arsenic	+	−	−	−	(+)	+	−	wY	−
Barium	−	−	−	+	−	−	−	R	
Beryllium	+	−	−	+	−	+	fl	−	−
Bismuth	+	+	+	+	+	−	−	Bl	Br
Cadmium	+	+	+	+	−	−	−	V	Bl
Caesium	−	−	−	+*	−	−	−	−	−
Calcium	+	−	−	+	−	−	−	R	−
Cerium	+	−	−	−	−	−	Br	Br	−
Chromium	+	−	+	+	+	−	−	Y	−
Cobalt	+	+	+	+	+	+	+	V	BrG
Copper	+	Br	+	+	−	+	+	BlV	Br
Gallium	+	−	−	−	−	+	−	−	−
Germanium	−	−	−	−	−	+	−	−	−
Gold	−	−	−	−	−	−	BlR	−	G
Hafnium	−	−	−	−	−	−	−	−	RP
Indium	+	−	−	−	−	−	−	−	−
Iron	+	−	Bk	+	+	+	R	wB	G
Lead	+	+	+	+	−	−	−	Bl	PGy
Lithium	−	−	−	+	−	−	−	RO	−
Magnesium	+	−	+	+	−	−	fl	R	−
Manganese	+	−	+	+	−	−	+	Br	Bl
Mercury	+	wPi	(II) +	+	−	−	−	(I) O	(I) GyBr
	−	−	(I)Bk	−	−	−	−	(II) Cr	(II) GGy
Molybdenum	−	−		−	+	+	−	−	−
Nickel	+	+	+	+	+	+	+	R	Bl
Potassium	−	−	−	+	−	−	−	−	−
Rubidium	−	−	−	+*	−	−	−	−	−
Silver	−	+	+	−	−	+	−	Gy	GyBr
Sodium	−	−	−	+*	−	−	−	−	−
Strontium	−	−	−	+	−	−	−	R	−
Thallium	−	wPi	+	−	−	−	−	R	wGy
Tin	−	(II)+	+	+	−	+	−	(II)GyBi	RBr
Thorium	+	−	−	−	+	+	−	−	−
Titanium	+	−	−	−	−	+	+	wY	Br
Tungsten	−	−	−	−	+	+	−	wBr	−
Uranium	+	−	+	−	−	+	+	GyBl	Br
Vanadium	−	−	−	−	+	+	−	(V) Och	−
								(IV) Br	
Zinc	+	+	+	+	−	−	fl	RV	R
Zirconium	+	−	−	−	+	+	−	V	RP
Rare earths	+	−	−	−	−	−	−	−	−

Notes: Roman numbers indicate the oxidation state concerned.

  * the colour does not appear immediately

acid and dried. Immediately afterwards, while still just damp, the chromatogram is exposed to furfurol vapour. The ions are detected as coloured spots on a blue-violet background. The method is suitable for the detection of silver, mercury, lead, bismuth, copper, cadmium, arsenic, antimony(V), tin(II), iron(III), aluminium, chromium, cobalt, nickel, zinc, manganese, barium, strontium, calcium, magnesium, sodium, potassium, lithium and ammonium. Lewandowski [3] used two methods for the detection of cations on paper chromatograms. The first detects cations forming insoluble precipitates with sodium carbonate, and the second is based on reactions with potassium ferrocyanide. In the first the dried chromatogram is immersed for 10 seconds in $1N$ sodium carbonate and then washed immediately with running water. The chromatogram is blotted between two filter papers and sprayed with a saturated sodium sulphate solution containing 0.1% of Bromothymol Blue, Phenol Red or Bromocresol Red. The spots immediately assume the colour of the alkaline form of the indicator or at least a colour intermediate between the acid and the alkaline form. This method of detection is most suitable for calcium, magnesium, strontium and barium, but various other elements also give sufficiently sensitive reactions (as little as 3 µg in a spot 5 mm in diameter). If sodium carbonate is not washed completely from the paper or the surplus water not removed, detection is impaired. Acetic acid can be used to whiten the background. The second method consists of dipping the chromatogram in $0.1M$ potassium ferrocyanide for 10 seconds, thorough washing with running distilled water and immersion

---

— negative reaction
+ positive reaction
(+) weak positive reaction

Colours: Bk — black, Bl — blue, Br — brown, G — green, Gy — grey, O — orange, Och — ochre, P — purple, Pi — pink, R — red, V — violet, Y — yellow; w — weak, fl — fluorescence.

Reagents and their application:

1 Alizarin — saturated solution in 96% ethanol. After spraying and drying the chromatogram is exposed to ammonia vapour. Red-violet spots appear. The background can be improved by heating or treatment with acetic acid.
2 Dithizone — 0.05% solution in carbon tetrachloride. Red-violet spots only.
3 8-Hydroxyquinoline — 0.5% solution in 60% ethanol. Yellow-green and yellow-brown hues. In many instances, the spots are better visible after treatment with ammonia vapour and observation under ultraviolet light.
4 Tetrahydroxy-*p*-benzoquinone — 0.2 g dissolved in 100 ml of ethanol. Exposure to ammonia vapour, red to brown spots.
5 Catechol Violet — 0.05% ethanolic solution. The reaction is strongly dependent on pH. In weakly acid solutions, the majority of cations gives blue or purple spots on a yellow background. In alkaline medium, some cations give green spots on a blue background (cobalt, nickel).
6 Quercetin — 1% ethanolic solution. Generally yellow to yellow-brown spots. Detection is often better when the chromatogram is observed under ultraviolet light.
7 2-Furoyltrifluoroacetone — 10% solution in 95% ethanol. Yellow to green colour.
8 Glyoxal-bis(2-hydroxyanil) — 1% methanolic solution. Spraying with the reagent, followed by spraying with a 3% sodium hydroxide solution.
9 Sodium pentacyanoamminoferrate [13] — 0.70 g of the reagent, formed on reaction of sodium nitroprusside with ammonia [14], is dissolved in 20 ml of water, a solution of 0.25 g of rubeanic acid in 10 ml of ethanol is added, and the mixture is filtered. After spraying with this reagent the chromatogram is washed with $0.2N$ acetic acid.

in $1N$ iron(II) chloride or copper(II) sulphate for 20 seconds. The spots of iron, copper, uranium and molybdenum assume a colour in the first bath, the rest only after immersion in the second. Distinct colours are given by silver, bismuth, cadmium, tin, molybdenum, vanadium, nickel, cobalt, manganese, zinc, thorium and uranium. The method cannot be used for the detection of arsenic, chromium, aluminium, beryllium, cerium, alkaline earth metals, and alkali metals.

Very often reagents are used which give intensely fluorescent derivatives with the substances sought. Under ultraviolet light (Wood's glass filter) minute amounts can be detected without destroying the chromatogram, by placing it on a suitable photosensitive material and exposing for an appropriate time interval to ultraviolet light [4, 5]. Light spots on a black background show the positions of the separated components.

Among reagents which give characteristic colours with several ions the following have been found useful: alizarin, Alizarin Red S [6], benzidine, dithizone, 8-hydroxyquinoline, tetrahydroxy-*p*-benzoquinone [7], Catechol Violet [8], quercetin [9], 2-furoyltrifluoroacetone [10], glyoxal-bis-(2-hydroxyanil) [11], and Rhodamine B [12]. The results obtained with some of these reagents are given in Table 4.1.

In addition to those listed in Table 4.1, other more specific reagents can be used [15]. The most useful of them are listed in Table 4.2.

For the detection of anions less specific reagents are preferable. Pollard and co-workers [22] proposed detection with a solution of silver nitrate and fluorescein, which was convenient for numerous anions (Table 4.3).

The detection of chloride has been used as an indirect test for some metal ions, especially alkali metal and alkaline earth metal chlorides. Silver nitrate is used, either mixed with fluorescein [23], or by first spraying the chromatogram with silver nitrate alone, washing off the excess with water, exposing the chromatogram to light, and developing it in a photographic metol-hydroquinone developer [24]. Ćelap [25] proposed detection by treatment of the well-washed chromatogram with ammonium sulphide.

4.1.2. Physical Methods

As physical methods of detection need more complicated apparatus, they are less often used. However, they are usually very sensitive, technically simple and especially suitable for the location of spots before quantitative determination. Detection is non-destructive, and the ion detected remains in a form amenable to quantitative chemical analysis.

Blake [26] detected separated ions on paper chromatograms by means of an electrostatic discharge. An electroscope was connected to a conducting surface on an electric lamp run from a regulated power supply. The chromatogram was drawn between two wire electrodes one of which was connected to the electroscope and the other to earth. When a conducting spot came between the electrodes the leaves of the electroscope moved. Instead of an electroscope a suitably connected neon tube could be used, in which case the tube glowed when 'blank' chromatogram passed between the electrodes but not when a spot was passing the electrodes.

Table 4.2. Detection of Cations

Element	Reagent	Colour	Remarks
Aluminium	1	R	The chromatogram is first sprayed with $1N$ HCl, then with the reagent, and eventually with ammonia
Antimony	50	O	After dissolving in HCl and spraying with phosphomolybdic acid — blue colour
	28	Y	—
	48	V	From a medium strongly acidified with hydrochloric acid
Arsenic	50	Y	—
	23	d	As^{3+} — after heating
Barium	22	Y	—
	25	P	—
	33	fl	Increased intensity after treatment with ammonia vapour
	45	P	—
	35	R	After heating at 60 °C
	49	R	After spraying with a mixture of $2M$ HCl and $2M$ acetic acid it does not decolorize
	30	Y—BrY	—
Beryllium	21	Bl	Violet background. The detection is not very distinct
	35	YG	After heating at 60 °C
	36	fl	—
	14	BlV	—
	1	R	—
Bismuth	50	Br	—
	22	Y	—
	28	O	—
	47	Y	—
	34	YBr	After exposure to ammonia vapour
	51	Y	—
	20	O	Reaction of iodide: more selective than inorganic iodide
	53	Y	—

Table 4.2 continued

Element	Reagent	Colour	Remarks
Cadmium	50	Y	—
	34	Y	After exposure to ammonia vapour
	39	Br	—
	40	Bl	—
	35	R	After heating at 60 °C
	30	Br	Washing with ether after detection. In ultraviolet light the spot is dark
	8	V	—
	9	V	—
Caesium	10	R	A number of elements interfere with the detection if the separation is not good
	31	wR	—
Calcium	33	fl	Increased intensity in ammonia vapour
	25	P	—
	45	P	—
	40	Bl	—
	35	O	After heating at 60 °C
	42	RV	After 10 minutes heating at 80 °C
	30	Y—BrY	—
Chromium	44a	Y	—
	44b	Bl	Low sensitivity and stability, but selective
	47	RP	—
	32	R	CrO_4^{2-}, disappears after spraying with dilute sulphuric acid
	1	R	—
	8	V	—
Cobalt	44a	Bk	—
	50	Bk	—
	47	Bl	Acetone in the reagent increases the intensity of coloration
	15	G	—
	34	Br	After exposure to ammonia vapour
	35	G	After heating at 60 °C
	26	dY	The colour intensifies gradually
	41	R	—
	30	Y	Washing with ether after detection. In ultraviolet light the spot is dark
	8	V	—
	9	V	—

Table 4.2 continued

Element	Reagent	Colour	Remarks
Copper	50	Br	—
	22	BrY	—
	15	RBr	—
	27	BrR	Insoluble in thiosulphate
	28	Br	—
	47	BrG—Bk	—
	12	R	Decoloration with a mixture of HCl and NH_4Cl. Does not decolorize with 5% KCN solution
	7	BrR	—
	34	G	After exposure to ammonia vapour
	35	YBr	After heating at 60 °C
	3	G	—
	30	wG	Washing with ether after detection. In ultraviolet light the spot is dark
	8	V	—
	9	V	—
Gallium	18	Bl	—
Germanium	17	Pi	—
Gold	50	Br	—
	47	O	—
	23	d	More sensitive after addition of iodide
	12	R	Decolorizes with 5% KCN solution
Iron	44a	O	—
	44b	Y	—
	15	Bl	Fluoride and phosphate prevent the reaction
	27	BrR	Insoluble in thiosulphate
	35	Bl	After heating at 60 °C
	4	RBr	—
	13	R	—
	41	G	—
	1	R	—
	24	O	—
	29	P	0.1 µg of iron can be detected
	30	Br	Washing with ether after detection. In ultraviolet light the spot is dark.
	47	R	After spraying with $SnCl_2$ solution in $2M$ HCl the spot disappears.
	32	G	Fe^{3+}; disappears after spraying with dilute H_2SO_4 or $SnCl_2$

Table 4.2 continued

Element	Reagent	Colour	Remarks
Iron	7	R	Fe^{2+} (under the influence of atmospheric oxygen the iron is usually present as Fe^{3+})
	34	BrBk	After exposure to ammonia vapours
Lead	50	Br	—
	22	Y	—
	27	Y	Soluble in thiosulphate and in excess of iodide
	28	Y	—
	34	Y	After exposure to ammonia vapour
	49	dV	—
	30	Br	Washing with ether after detection. In ultraviolet light the spot is dark
	8	V	—
	9	V	
Lithium	43	Y	Green fluorescence in ultraviolet light, intensified with acetic acid, quenched in ammonia vapour
	35	RV	After heating at 60 °C
	42	RV	After 10 minutes heating at 80 °C
	30	V	At higher concentration almost brown-yellow
Magnesium	25	P	—
	45	P	—
	39	RV	In alkaline medium. Not very suitable for detection, dependent on the concentration of OH^-
	40	Bl	In alkaline medium (ammonia or 1% NaOH)
	35	YPi	After heating at 60 °C
	42	RV	After 10 minutes heating at 80 °C
	30	YbrY	—
	8	V	—
Manganese	44a	dBr	—
	34	YBr	After exposure to ammonia vapour
Mercury	44a	Bk	—
	50	Bk	Insoluble in conc. HCl
	22	Y	Hg^{2+}; Hg_2^{2+} red-orange
	27	YG	Soluble in thiosulphate and excess of iodide

Table 4.2 continued

Element	Reagent	Colour	Remarks
Mercury	28	Y	Hg^{2+}
	47	Bk	Hg_2^{2+}
	23	d	Hg_2^{2+} in the cold, Hg^{2+} after heating
	32	Y	Disappears after spraying with dilute sulphuric acid
	12	R	Decolorizes with 5% KCN solution
	33	fl	Intensified in ammonia vapour
	34	YG	Hg^{2+}; after exposure to ammonia vapour
		Bk	Hg_2^{2+}; after exposure to ammonia vapour
	30	Br	Hg^{2+}; washing with ether after detection. In ultraviolet light the spot is dark
	8	V	—
	9	V	—
Molybdenum	15	dBr	—
	47	R	Does not disappear after spraying with $SnCl_2$ in $2M$ HCl
	16	Y	After reduction with $SnCl_2$ it turns blue
Nickel	44a	Bk	—
	50	Bk	—
	47	G	—
	7	R	In neutral medium or in acetic acid or ammonia. Oxidizing substances should not be present
	34	Bl	After exposure to ammonia vapour
	30	G	Washing with ether after detection. In ultraviolet light the spot is dark
	8	V	—
	9	V	—
Niobium	15	YBr	—
Phosphorus	37	Y	After detection the chromatogram is exposed to H_2S or observed under ultraviolet light
Potassium	5	Y	At higher concentrations green to black. Sulphates interfere (if present, spray with saturated barium nitrate solution first)

Table 4.2 continued

Element	Reagent	Colour	Remarks
	10	O	A number of elements interfere if not separated
	35	V	After heating at 60 °C
	42	RV	After 10 minutes heating at 80 °C
	30	V	Almost brown-yellow at higher concentrations
Rubidium	10	R	A number of elements interfere if not separated
Selenium	25	R	Reduction
Silicon	38	Y	With molybdate, after heating on a water-bath
		Bl	With molybdate, then with benzidine — exposure to ammonia vapour
Silver	44a	Bk	—
	50	Bk	—
	22	R	Disappears with ammonia, turns black on irradiation
	27	Y	Soluble in thiosulphate
	28	Y	—
	23	d	In the cold
	32	Br	Turns black with $SnCl_2$
	12	R	Does not disappear under influence of a mixture of HCl and NH_4Cl, or 5% KCN solution
	33	fl	More distinct after exposure to ammonia vapour
	34	BrBk	After exposure to ammonia vapour
	30	BrR	Washing with ether after detection. In ultraviolet light the spot is dark
	8	V	—
	36	BlR	—
Sodium	43	Y	Under ultraviolet light characteristic green fluorescence, intensified with acetic acid, quenched with ammonia
	35	V	After heating at 60 °C
	42	RV	After 10 minutes heating at 80 °C
	30	V	At higher concentration almost yellow-brown

Table 4.2. continued

Element	Reagent	Colour	Remarks
Strontium	25	P	—
	45	P	—
	35	RV	After heating at 60 °C
	49	wR	Decolorizes with a mixture of $2M$ HCl and $2M$ acetic acid
	30	Y—BrY	—
Tellurium	23	d	Reduction
Thallium	44a	dBr	—
	50	Br	—
	22	Y	Tl^+
	27	Y	Does not change with thiosulphate
	28	Y	—
	10	R	A number of elements interfere if not separated
	31	Br	—
Tin	50	Br	—
	9	V	—
Titanium	44a	O	—
	44b	Y	—
	15	YBr	—
	16	Y	Does not change after reduction with $SnCl_2$
	32	BrR	Does not disappear after spraying with dilute sulphuric acid
Tungsten	15	Y	—
Uranium	44a	Y	—
	15	Br	—
	33	Y	—
	22	Br	Disappears after spraying with dilute H_2SO_4 or $SnCl_2$
	30	Br	Washing with ether after detection. In ultraviolet light the spot is dark
Vanadium	44a	O	—
	44b	BrR	Chromium interferes if not separated
	13	YG	—
	50	Br	—
	22	RO	—
	28	Y	—

Table 4.2 continued

Element	Reagent	Colour	Remarks
Zinc	35	R	After heating at 60 °C
	30	G	Washing with ether after detection. In ultraviolet light the spot is dark
Zirconium	18	BlG	—
Platinum metals	50	Bk	Rhodium
		Br	Palladium, platinum
	47	R	Platinum
	23	d	Platinum, palladium, rhodium. More sensitive if potassium iodide is added to the reagent
	12	R	Platinum, palladium. Decolorized with 5% KCN solution

Notes:

Colours: Bk — black, Bl — blue, Br — brown, G — green, O — orange, P — purple, Pi — pink, R — red, Y — yellow; w — weak, d — dark, fl — fluorescence

Reagents:

1 Aluminon — 0.1% of Aluminon and 1% of ammonium acetate in water. The colour appears after exposure of the chromatogram to ammonia vapour.
2 Benzidine — 0.5 g of benzidine dissolved in 10 ml of glacial acetic acid, diluted to 100 ml with water and filtered.
3 α-Benzoinoxime — 1% ethanolic solution.
4 Dibenzoylmethane — 1% ethanolic solution.
5 Cobaltinitrite and lead nitrite — a few drops of $2M$ acetic acid are added to a 5% aqueous solution of sodium cobaltinitrite and 5% lead nitrite.
6 Silver nitrate — $2M$ aqueous solution.
7 Dimethylglyoxime — 1% ethanolic solution.
8 Diphenylcarbazide — 1% ethanolic solution.
9 Diphenylcarbazone — saturated ethanolic solution.
10 Dipicrylamine — 1% aqueous solution.
11 Dithizone — freshly prepared 0.05% solution in carbon tetrachloride.
12 *p*-Dimethylaminobenzylidenerhodanine — 0.03% ethanolic solution.
13 α, α'-Dipyridyl — 1% ethanolic solution.
14 Eriochromecyanine R — 0.07% aqueous solution.
15 Potassium ferrocyanide — 5% aqueous solution.
16 Sodium phosphate and stannous chloride — $2M$ aqueous sodium phosphate is acidified with conc. nitric acid — 3% stannous chloride in $2M$ hydrochloric acid.
17 Phenylfluorone — 0.5% ethanolic solution acidified with hydrochloric acid.
18 Formalhydrazone of resorcylaldehyde — 0.01% solution in redistilled alcohol.
19 8-Hydroxyquinoline — 0.5 g dissolved in 100 ml of 60% ethanol.
20 Hexamethylenetetramie-allyl iodide — 1% ethanolic solution.
21 Quinalizarin — 0.05 g of reagent dissolved in 100 ml of $0.1M$ sodium hydroxide.
22 Potassium chromate — 1% aqueous solution.
23 Stannous chloride — 5% solution in $5M$ hydrochloric acid.
24 1,10-Phenanthroline — 0.2% ethanolic solution.
25 Gallacetophenone [16] — 1% aqueous solution.

Oehme [27] modified the method of detection proposed by Hashimoto [28] based on perturbing a high-frequency oscillator circuit, and described a device for recording high-frequency detection. An electrolyte passing between the plates of a condenser caused a change in high-frequency conductance which was recorded by measuring the grid current of a quartz oscillator. It was possible to obtain semiquantitative results with this apparatus when a well-defined working procedure was adhered to and when the humidity of the chromatogram was kept constant. De Vries [29] proposed a similar arrangement based on conductimetric detection of ions on a paper chromatogram. After drying, the paper strip was moved at a constant rate between two narrow steel cylinders having a constant d.c. potential difference of 4—80 V. At 3-second intervals, corresponding to 0.75 mm travel, the current was read on a spot galvanometer and the results were evaluated graphically. De Vries and Dalen proposed a similar device [30] in which

---

Table 4.2 continued

26 Isonitrosoacetophenone — 0.5% ethanolic solution.
27 Potassium iodide — 2% aqueous solution.
28 Potassium iodide and cinchonine — 1 g of cinchonine dissolved in 100 ml of hot water, acidified with a few drops of conc. nitric acid. After brief boiling, 2 g of potassium iodide are added.
29 2,4,6-Pyridinetricarboxylic acid — 0.1% solution in methanol containing 5% of pyridine.
30 Chloranilic acid [17] — 0.1% ethereal solution.
31 Bismuth potassium iodide — 1 g of bismuth oxide is dissolved in boiling saturated KI (5g) solution and 25 ml of acetic acid added dropwise.
32 Chromotropic acid — 5% aqueous solution.
33 Kojic acid [18, 19] — 0.1% aqueous solution.
34 Rubeanic acid — 0.5% ethanolic solution.
35 Violuric acid [10] — 0.1% aqueous solution.
36 Morin — 0.02% acetone solution.
37 Ammonium molybdate — an aqueous solution acidified with hydrochloric and perchloric acids.
38 Ammonium molybdate — 5 g of ammonium molybdate in 100 ml of water poured into 35 ml of nitric acid (sp. gr. 1.2); 0.05% of benzidine in glacial acetic acid.
39 Magneson I — 0.01% solution in 50% aqueous ethanol.
40 Magneson II — 0.01% ethanolic solution.
41 Nitroso-R-salt — 0.01% solution in a mixture of water and acetone (70 : 30).
42 Ninhydrin — 0.1% ethanolic solution.
43 Zinc uranyl acetate — saturated solution in $2M$ acetic acid (filtered).
44 Hydrogen peroxide — (a) alkaline — a mixture of equal amounts of $2M$ sodium hydroxide or ammonia (sp. gr. 0.88) and 30% hydrogen peroxide; (b) acid — a mixture of equal amounts of $2M$ sulphuric acid and 30% hydrogen peroxide.
45 Pyrogallol-4-carboxylic acid [2] — 1% solution in 50% ethanol.
46 Catechol Violet — 0.05% ethanolic solution.
47 Potassium thiocyanate — aqueous solution, or a mixture of equal amounts of a saturated aqueous solution of potassium thiocyanate or ammonium thiocyanate and acetone.
48 Rhodamine B — 0.01% aqueous solution.
49 Sodium rhodizonate — 0.1% aqueous solution.
50 Yellow ammonium sulphide — a measured amount of ammonia is saturated with gaseous hydrogen sulphide, diluted with an equal amount of concentrated ammonia, and sulphur is dissolved in this solution.
51 Thioacetamide — 2% solution in $1M$ hydrochloric acid.
52 Thymol Blue (thymolsulphonphthalein) — 0.1% ethanolic solution.
53 Thiourea — 2% solution in $1M$ hydrochloric acid.

Table 4.3. Detection of Anions

Anion	Cation	Colour	
		visible light	ultraviolet light
Fluoride	Sodium	dBr*	wY-fl
Chloride	Potassium	G	Y-fl
Bromide	Potassium	G	Y-fl
Iodide	Potassium	Y	YG-fl
Bromate	Potassium	—	YG-fl
Iodate	Potassium	—	wY-fl
Periodate	Potassium	—	—
Sulphite	Sodium	Y	Y-fl
Sulphate	Ammonium	wBr	G
Thiosulphate	Sodium	G	dBl
Dithionate	Sodium	dR*	—
Trithionate	Potassium	G	dBl
Disulphite	Sodium	—	wG
Nitrite	Sodium	Y	Y-fl
Borate	Sodium	Y-R*	Y-fl
Carbonate	Sodium	G, turns slowly R	GY-fl
Thiocyanate	Ammonium	Y	Y-fl
Ferrocyanide	Potassium	Y	Y-fl
Ferricyanide	Potassium	—	d
Formate	Potassium	dBr	d
Acetate	Sodium	wR	—
Oxalate	Ammonium	wBr	Y-fl
Tartrate	Potassium	Y	Y-fl
Citrate	Sodium	Y	Y-fl
Phthalate	Potassium (KH)	—	d
Molybdate	Ammonium	wY	wG-fl
Vanadate	Ammonium	Y	G
Hydroxide	Sodium	G	YG-fl
Bicarbonate	Potassium	—	G-fl
Lactate	Ammonium	wBr	wG
Arsenate	Sodium	wBr	dG
Arsenite	Sodium	wBr	dG
Dihydrogen phosphate	Sodium (NaH$_2$)	BlG	Bl
Hydrophosphate	Sodium (NaH$_2$)	G	YG-fl
Diphosphate	Sodium	Y	Y-fl
Phosphate	Sodium	BrR	—
Trimetaphosphate	Sodium	BrR	—
Tetrametaphosphate	Sodium	BrR	—
Monothiophosphate	Sodium	G	d
Dithiophosphate	Sodium	G	d
Trithiophosphate	Sodium	Y	d
Imidodisulphonate	Ammonium	Y	Y-fl
Amidosulphonate	Ammonium	Y	Y-fl
Sulphamide	—	Y	Y-fl

the dried chromatogram was drawn between two steel cylinders, the upper of which carried a 1.5 kg load and the lower was 1.5 cm shorter than the width of the chromatogram (to avoid a short circuit).

Langer [31] described an apparatus in which ions were detected polarographically at a gold wire electrode making contact with the paper passing over a porous porcelain cylinder dipped in a saturated potassium chloride solution. Pechan and co-workers [32] made use of a stationary amalgam-

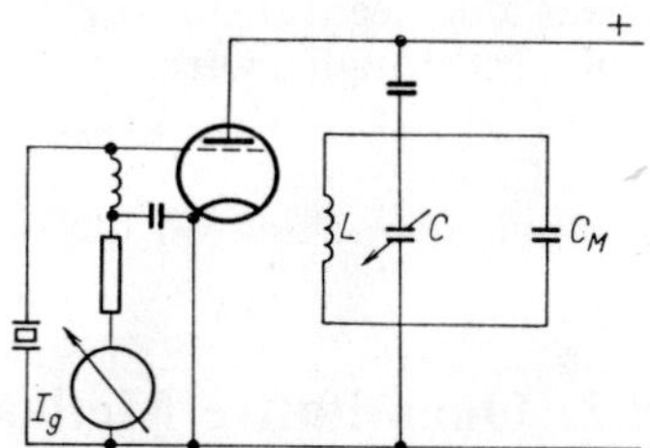

Fig. 4.1. Circuit diagram of a simple quartz oscillator for high-frequency registration paper chromatography.

coated copper electrode and an oscillopolarographic detection of polarographically active substances.

For the determination of purity of the separated spots on chromatograms and for a rapid rough evaluation of the composition of fractions in eluates from column chromatography, spectral analysis may be used. Burstall and co-workers [33] applied this technique to control of the separation of niobium from tantalum and other metals, and Spedding [34] used it for checking the purity of chromatographically separated rare earths.

Spectrophotometric analysis is generally only used for substances absorbing in the ultraviolet since coloured substances may be detected by eye. On the other hand, spectrophotometry is often used for the quantitative determination of separated components in eluate fractions from columns.

Radiochemical techniques are often associated with chromatographic separations, and may be divided into two groups. The first deals with the addition of labelled compounds to the mixture to be separated, and the second is concerned with activation methods of analysis.

For the detection of radioactive substances on chromatograms either autoradiography or scanning radio-counting may be used. In the former the chromatogram is put into close contact with a suitable photographic material (depending on the radioactivity of the labelled compounds used). It is advisable to determine the optimum time necessary to obtain a suitable negative, by preliminary experiments. If radiation other than soft β-radiation is to be detected a cellophane film may be positioned between the chromato-

---

Notes to Table 4.3:

Colours: Bl — blue, Br — brown, G — green, R — red, Y — yellow; w — weak, d — dark, fl — fluorescence

* After exposure to ultraviolet light

Preparation of the reagent: One part of 10% aqueous silver nitrate solution is mixed with 5 parts of a 0.2% alcoholic sodium fluorescein solution.

gram and the film to prevent haziness due to chemical activity of substances on the paper. A suitable arrangement for photographic recording of auto-radiograms was described by Horešovský and co-workers [35] and a very sensitive apparatus, applicable to thin-layer chromatography, has been proposed by Fray and Frey [36].

Radio-counting using a Geiger-Müller tube or an ionization chamber offers direct measurement of activity. A simple measuring apparatus has been described by Linskens [37]. He inserted into a Philips Universal lead tower a device transmitting the radiation from the paper to the Geiger-Müller counter through a narrow slit. The chromatogram was moved slowly through the lower part of the tower. This device required manual control and record-ing of count-rates, but it is not difficult to make fully automatic recording equipment working on the same principle.

4.2. Quantitative Methods

In column chromatography the eluate is analysed continuously or as fractions of small volume, and in both cases the solutions are usually rather dilute with the result that the analyst is compelled to utilize methods common in semimicro- or microanalysis. The same situation is met after dissolution of components separated by paper or thin-layer chromatography.

In paper chromatography one of the main problems affecting the success of the whole analysis is the accurate measurement of the volume of the solution to be analysed. Pipettes with a micrometer scale, which allow a very accurate measurement, are available (see Section 3.2.3), but even so the use of paper chromatography for the accurate determination of macrocompo-nents in inorganic analysis is often problematical. In some cases a normali-zation procedure can be chosen, making accurate volume measurement unnecessary. Fidler and Michal [39] used the following procedure for the analysis of rare earths: first they determined the total rare-earth content gravimetrically, then separated the rare earths chromatographically and determined each one separately photometrically. From the ratios of these values they then calculated the percentage of each component.

There are several approaches to quantitative analysis which do not depend on elution of the substance, but involve instead measurement of spot-size, by photometry, retention analysis, conductivity, polarography, etc.

The simplest, though relatively the least accurate approach is that based on the size of the spots measured either planimetrically or by drawing the outline of the spot on a piece of paper, cutting it out and weighing it. The relationship between the logarithm of the amount of the substance in the spot and the area of the spot is usually linear. In many publications other quantitative relationships have been described, e.g. between the logarithm of the quantity and the logarithm of the length of the spot, and between the length of the spot and the logarithm of the amount [39]. It is therefore usu-ally necessary to assess the relation experimentally by means of a calibration chromatogram before the quantitative determination. As the absolute amount of the chromatographed substance does not itself decide the definite shape and size of the spot, but also the volume of solution applied, the diameter of the spot at the start and a series of other factors, it is often necessary to chromatograph control samples (standards) in parallel with the analysed sample, or, even better, to use the internal-standard method (addition of standard to the analysed sample).

In semiquantitative analysis or the chromatography of components at very low levels, determination simply by comparison with a calibrated chromatogram may be adequate, particularly if detection can be carried out with sensitive reagents giving pronounced and sharp spots.

Some physico-chemical methods are more accurate than planimetric methods. The high-frequency oscillator and polarographic methods have been mentioned already in Section 4.1, but the most convenient and widely used technique is spectrophotometry, either in the visible or the ultraviolet.

As the precision of photometric analysis is influenced by the nature of the medium, chemical equilibria, losses during chromatography, and many other factors, it is necessary as a rule to work under constant conditions, i.e. always to work with the same volumes of substances, to chromatograph at a constant rate, and to chromatograph standards in parallel with a series of samples, which allows the construction of a calibration curve. For absorbance measurements on paper it is advantageous to increase the transparency of the paper by impregnation with a substance having approximately the same refractive index as cellulose (for example paraffin oil).

For photometric measurements a number of instruments have been developed, known as densitometers, which carry out all the necessary operations under optimum conditions, including the measurement of the intensity of the transmitted or reflected light, automatic calculation of logarithms, measurement of any backing-off potential, and the final integration. Thus the apparatus is calibrated directly. Such densitometers have been developed by Mastner and co-workers [40], Halama [41], Príbela [42], Rendoš [43], Pavlů [44] and others.

Jackwerth and Kloppenburg [45, 46] proposed a method of evaluation of paper chromatograms based on X-ray fluorescence spectroscopy. For recording they used two methods; planimetric recording of the count-rate vs. R_F diagrams, and measurement of fluorescent radiation by the determination of count-rate on a reference chromatogram. With measuring times from 5 to 15 minutes they achieved, on repeated measurement of the same chromatogram, reproducibility to within $\pm 1\%$, but on measurement of equal amounts of analysed substances on different chromatograms the errors increased to 2.5%. For the determination of zinc and gallium they obtained good results even when the amount determined was less than 1 µg.

Yamada [47] constructed a coulometer for quantitative analysis. He was able to determine chromatographically separated copper in amounts from 1 to 300 µg within $\pm 5\%$.

REFERENCES

1. FEIGL, F. *Spot Tests Vol. I. Inorganic Applications*, Elsevier, New York (1954).
2. GÁNTI, T. and NOVAK, E. *Magy. Kem. Folyoirat*, **68**, 293 (1962).
3. LEWANDOWSKI, A. *Chem. Analit.* (*Warsaw*), **4**, 539 (1959).
4. ABELSON, D. M. and ROSENFELD, L. *J. Chromatog.* **9**, 522 (1962).

5. MILTON, A. S. *J. Chromatog.* **8**, 417 (1962).
6. HEISIG, G. B. and POLLARD, F. H. *Anal. Chim. Acta* **16**, 234 (1957).
7. BOCK-WERTHMANN, W. *Anal. Chim. Acta* **28**, 519 (1963).
8. MACEK, K. and MORÁVEK, L. *Nature* **178**, 102 (1956).
9. MICHAL, J. *Chem. Listy* **50**, 77 (1956); *Collection Czech. Chem. Commun.* **21**, 576 (1956).
10. MCINTYRE, R. T., BERG, E. W. and CAMPBELL, D. N. *Anal. Chem.* **28**, 1316 (1956).
11. MÖLLER, H. G. and ZELLER, N. *J. Chromatog.* **14**, 560 (1964).
12. MIKEŤUKOVÁ, V. *J. Chromatog.* **24**, 302 (1966).
13. HARROP, D. and HERINGTON, E. F. G. *Analyst* **81**, 499 (1956).
14. HERINGTON, E. F. G. *Analyst* **78**, 174 (1953).
15. BOCK-WERTHMANN, W. *Z. Anal. Chem.* **198**, 403 (1963).
16. POLLARD, F. H., MCOMIE, J. F. W. and STEVENS, H. M. *J. Chem. Soc.* 1863 (1951).
17. BARRETO, H. S., BARRETO, R. C. R. and PINTO, I. P. *J. Chromatog.* **5**, 5 (1961).
18. POLLARD, F. H., MCOMIE, J. F. W. and ELBEIH, I. I. M. *Nature* **163**, 292 (1949).
19. ELBEIH, I. I. M., MCOMIE, J. F. W. and POLLARD, F. H., *Disc. Faraday Soc.* **7**, 183 (1949).
20. GUINCHARD, J. *Ber.* **32**, 1739 (1899).
21. OKÁČ, A. and PECH, J. *Collection Czech. Chem. Commun.* **13**, 400, 514 (1948).
22. POLLARD, F. H., NICKLESS, G. and BURTON, K. W. C. *J. Chromatog.* **8**, 507 (1962).
23. RUDOLFF, H. *Deut. Apotheker-Ztg.* **102**, 1571 (1962).
24. MAZZA, L. and SARDO, L. *Gazz. Chim. Ital.* **93**, 1155 (1963).
25. ĆELAP, M. B. *Glasnik Chem. Drustva, Belgrade* **21**, 225 (1956).
26. BLAKE, G. G. *Anal. Chim. Acta* **17**, 489, 492 (1957).
27. OEHME, F. *Z. Anal. Chem.* **150**, 93 (1956).
28. HASHIMOTO, A. and MORI, I. *Nature* **172**, 542 (1953).
29. DE VRIES, G. *Nature* **173**, 735 (1954).
30. DE VRIES, G. and VAN DALEN, E. *Rec. Trav. Chim.* **73**, 1028 (1954).
31. LANGER, A. *Anal. Chem.* **28**, 426 (1956).
32. PECHAN, Z., KALÁB, D. and PALEČEK, E. *Pharmacie* **10**, 526 (1955).
33. BURSTALL, F. H., SWAIN, P., WILLIAMS, A. F. and WOOD, G. A., *J. Chem. Soc.* 1497 (1952).
34. SPEDDING, F. H., *Disc. Faraday Soc.* **7**, 214 (1949).
35. HOREŠOVSKÝ, O., DUPAL, J. and HAIS, I. M. *J. Chromatog.* **9**, 260 (1962).
36. FRAY, G. and FREY, J. *Bull. Soc. Chim. Biol.* **45**, 1201 (1963).
37. LINSKENS, H. F. *J. Chromatog.* **1**, 471 (1958).
38. FIDLER, J. and MICHAL, J. *Rudy* **11** (No. 10), supplement No. 5, 1 (1963).
39. FOWLER, H. D. *Nature* **168**, 1123 (1951).
40. MASTNER, J., FRANĚK, F. and NOVÁK, L. *Collection Czech. Chem. Commun.* **24**, 2959 (1959).
41. HALAMA, D. *Chem. Zvesti* **13**, 254 (1959).
42. PRÍBELA, A. *Chem. Zvesti* **17**, 689, 816 (1963).
43. RENDOŠ, F. *Chem. Zvesti* **17**, 916 (1963).
44. PAVLŮ, J. *Chem. Listy* **58**, 309 (1964).
45. JACKWERTH, E. and KLOPPENBURG, H. G. *Naturwiss.* **47**, 444 (1960).
46. JACKWERTH, E. and KLOPPENBURG, H. G. *Z. Anal. Chem.* **179**, 186 (1961).
47. YAMADA, T. *Bunseki Kagaku,* **3**, 215 (1954).

Chapter 5

SYSTEMATIC QUALITATIVE ANALYSIS

Chromatographic separations lend themselves readily to systematic qualitative analysis of cations and anions. Some procedures are based only on chromatographic separation, while others make use of selective reactions with various analytical reagents after partial chromatographic separation. All the procedures, especially those requiring a preliminary separation or chemical pretreatment, make use of typical microchemical methods of analysis to handle the separated elements in the 2 − 100 µg range. If only trace components in the sample are sought, the volume should be kept to a minimum in order to keep the concentrations of trace elements above the limits of detection.

5.1. Systems of Qualitative Analysis of Cations

A number of procedures have been worked out for the qualitative analysis of cations, some of which deal only with particular problems, such as the separation of a given group of elements. Generally, these partial procedures, if combined systematically so as to cover the whole range of cations, could yield a complete qualitative analysis. However, it is better to choose one of the systems for complete analysis of an unknown sample. The procedures are usually based on one of the following three principles.

1 − The cations are first separated into groups by chemical reactions with group-specific reagents and chromatography is applied only for separations within these groups (Section 5.1.1).
2 − The cations are partly separated by chromatography and then detected by selective or specific reactions (Section 5.1.2).
3 − The cations are separated in a series of solvent systems and are then determined predominantly on the basis of the positions of the separated spots (Section 5.1.3).

In this chapter some systems based on these principles are described. Only the most important systems have been chosen from those published. The reader can find other systems in the original literature [1 − 7].

5.1.1. Chromatographic Analysis of Chemically Separated Groups of Cations

Procedures in which an unknown mixture of elements is first separated on the basis of suitable chemical reactions into groups are preferred by many authors. In these, chromatography is carried out with a small number of ions, and the results are simpler, clearer and often unambiguous. Best known are the procedures by Pfeil [8], Blasius [9], and Biswas [10].

Pfeil's System

This system [8] is based on the division of cations into the four analytical groups: insoluble sulphides (in acid and alkaline media), carbonates, and alkali metals.

For the hydrogen sulphide group (acidic medium) the chromatograms are run with 1-butanol saturated with $3.4M$ hydrochloric acid and detection is done with 0.2% potassium iodide solution (Hg — a red spot, turning yellow to colourless with an excess of the reagent; Bi — brown, yellow with excess; Sb, Cu — brown colour due to the liberated iodine; Pb — faint yellow), then by exposure to ammonia vapour and spraying with $0.1M$ ammonium sulphide (Hg — red; Cd — yellow; Bi — brown; Pb — black; Cu — brown-black; Te — black; Se — black); and finally by treatment with 0.1% alcoholic quercetin solution and exposure to ammonia (Sn — yellow; Ge — yellow).

Ions of the ammonium sulphide group (alkaline medium) are chromatographed with a mixture of acetic acid, pyridine, and hydrochloric acid (40 : 3 : 10) and detected with a reagent due to Eegriwa (equal parts of a saturated diethylaniline solution in 4% oxalic acid and a 4% potassium ferricyanide solution); (Fe — blue; Zn — red-brown), or with alkaline hydrogen peroxide (Mn — brown), ammonium sulphide (Ni, Co — black), dimethylglyoxime, morin, or other organic reagents.

In the ammonium carbonate group the ions are chromatographed with a mixture of methanol and hydrochloric acid, and detected by spraying with methanolic 8-hydroxyquinoline solution, drying, and treating with ammonia. The reagent gives spots with alkaline earth metals, which fluoresce strongly, emitting yellow or yellow-green light.

Alkali metals are separated with a mixture of methanol and acetic acid (98 : 2), but the ions must be applied on the chromatogram as their acetates. The optimum quantity is $20-30$ μg. Detection is best done by spraying the still wet chromatogram with a solution of Bromocresol Green: alkali metals appear as blue spots on a yellow background. R_F values of cations of the single groups are listed in Table 5.1.

Blasius's System

Blasius [9] proposed a systematic qualitative analysis of an unknown mixture of cations, based on their division into 10 groups. Each group is then separated by circular paper chromatography, four samples of the same

Table 5.1. R_F Values in Pfeil's System of Qualitative Chromatographic Analysis of Cations

Hydrogen sulphide group				Ammonium sulphide group				Ammonium carbonate group		Alkali metals	
ion	R_F	ion	R_F	ion	R_F	ion	R_F	ion	R_F	ion	R_F
Ag	0.0	Bi	0.65	Fe	0.92	VO_3^-	0.35	Be	0.95	Li	0.66
Cu	0.22	Se(IV)	0.73	Zn	0.82	Al	0.30	Mg	0.73	Na	0.48
Pb	0.32	As(V)	0.83	Co	0.68	Cr	0.26	Ca	0.43	K	0.32
Sb(V)	0.39	Cd	0.85	UO_2^{2+}	0.68	Ti(IV)	0.20	Sr	0.20	—	—
Ge	0.49	Sn(IV)	0.92	Mn	0.57	—	—	Ba	0.07	—	—
Te	0.56	Hg(II)	0.94	WO_4^{2-}	0.43	—	—	Ra	0.0	—	—
Mo	0.62	—	—	Ni	0.39	—	—	—	—	—	—

material being applied on the chromatogram. The start is a circle 1.5 cm in diameter and the lines of application lie at an angle of 45° to the direction of the paper fibres. The sample (approx. 50 μg) is dissolved as far as possible in 1−2ml of very dilute nitric acid and is completely used in the analysis.

The first of the ten groups is the hydrogen chloride group, comprising the ions of silver, lead, mercury(I), and thallium. After centrifuging the chlorides are dissolved in a mixture of ammonium acetate and thiocyanate, with warming if necessary, and then applied and chromatographed with a 10% solution of ammonium acetate. Detection is by spraying with ammonium sulphide. R_F values: Pb 0.95, Ag 0.80, Th 0.75, Hg(I) 0.65.

Any insoluble residue after dissolution in dilute nitric acid is dissolved in concentrated nitric acid, and the solution is combined with the filtrate from the chloride precipitation, and treated with hydrazine to precipitate the metals of the reduction group. After filtering off they are redissolved in concentrated nitric acid with added hydrogen peroxide and then chromatographed in a mixture of ethanol and $2M$ acetic acid (8 : 2). The elements have the following R_F values: Au 1.0, Se 0.60, Pt 0.50 and 0.90, Te 0.0, and are detected with a solution of stannous chloride and potassium iodide in hydrochloric acid.

The copper-arsenic group is next precipitated by addition of a hot 2% solution of thioacetamide. After washing, the sulphides of the arsenic group are dissolved by repeated 10-minute shaking with $0.5M$ potassium hydroxide, while the copper group sulphides remain undissolved.* The latter are washed first with water containing potassium hydroxide, then with hydrogen sulphide water, and dissolved in a mixture of hydrochloric acid and hydrogen peroxide. The copper group is separated with a mixture of t-butyl alcohol, acetone, water, acetylacetone, and nitric acid (49 : 33 : 10 : 5 : 3); the R_F

---

* For a discussion of this group separation see A. I. Vogel, W. T. Cresswell, A. H. Jeffery and J. Leicester, *Analyst*, **81**, 244, (1956) and references therein.

values of the separated elements are Hg 1.0, Bi 0.65, Cu 0.35, Cd 0.25, Pb 0.15. The arsenic group is separated with a mixture of ethyl acetate, methanol, and 10% aqueous oxalic acid solution (68 : 12 : 20); R_F values are Sb 1.0, Sn 0.80, As 0.50, Mo 0.40, Pt 0.20. For the detection of both groups dithizone, rhodizonic acid (for Pb), and iodide (for Sb, As) are used.

In the filtrate from the hydrogen sulphide group thioacetamide is decomposed by evaporation with nitric acid. At this stage of the analysis it is necessary to eliminate the interfering anions. Chromate and permanganate are reduced by the addition of ethanol, but phosphates and tungstates must be separated on ion-exchangers after the elimination of nitric acid. If 'rare' elements are absent from the solution a strongly acid cation-exchanger is suitable, as for example Permutit RS in the H^+-form; otherwise an anion-exchanger is required, as for example Permutit ESB. The eluate from the ion-exchange column is buffered with ammonia and ammonium carbonate, and hydroxides are precipitated by warming with hexamine and separated. They are dissolved in hydrochloric acid, and this solution is added dropwise to a mixture of equal parts of 30% sodium hydroxide solution and 3% hydrogen peroxide. On heating, hydroxides of the titanium group precipitate, while the chromium and aluminium group ions remain in solution. The hydroxides are separated by filtration, dissolved in hydrochloric acid and chromatographed with a mixture of ethanol and hydrochloric acid (83 : 17). After detection with 8-hydroxyquinoline, drying, and spraying with Alizarin S and ammonia, spots appear for zirconium R_F 0.0−0.25, cerium 0.25−0.15, titanium 0.55, and iron 1.0.

The filtrate from the hydroxide and peroxide precipitation is slightly acidified with hydrochloric acid, evaporated to 1 ml, saturated with sulphur dioxide, and boiled to eliminate the excess of the latter. After addition of hydrosulphite, hydroxides of the chromium group are precipitated by addition of sodium hydroxide (final NaOH concentration should be $2M$). The precipitate is dissolved in hydrochloric acid, chromatographed with a mixture of acetone, t-butyl alcohol, hydrochloric acid and acetylacetone (33 : 43 : 19 : 5), and detected with 8-hydroxyquinoline and ammonia (U, V), or with a solution of sodium peroxide and benzidine (Cr). $R_F$values are Cr 0.30, V 0.50, U 0.80.

The filtrate from the chromium group is neutralized with hydrochloric acid to precipitate aluminium and beryllium hydroxides. After filtration, washing, and dissolving in hydrochloric acid they are applied to the chromatogram and run in a mixture of ethyl methyl ketone and hydrochloric acid (85 : 15). Detection is carried out with 8-hydroxyquinoline and ammonia, R_F Al 0.30 and Be 0.55.

Sulphides of the zinc group are precipitated from the filtrate from hexamine precipitation by addition of ammonium sulphide, filtered, washed, dissolved in hydrochloric acid and hydrogen peroxide, and chromatographed in a mixture of ethyl methyl ketone and hydrochloric acid (91 : 9). Zinc is detected with dithizone (red spot, R_F 0.85). A reagent containing alizarin, rubeanic acid and salicylaldoxime is used for the detection of cobalt and manganese, brown spots of R_F 0.60 and 0.30 respectively, and nickel (blue spot, R_F 0.05).

The filtrate from the precipitation with ammonium sulphide is acidified with hydrochloric acid and evaporated. Addition of nitric acid and evapora-

Table 5.2. Scheme of Qualitative Chromatographic Analysis of Cations
According to Biswas and Dey

Lead, mercury (I), and silver must be separated in the form of chlorides and determined by conventional methods. The filtrate, containing cations of the groups II, III, IV and V, is saturated with hydrogen sulphide and the precipitate is separated.

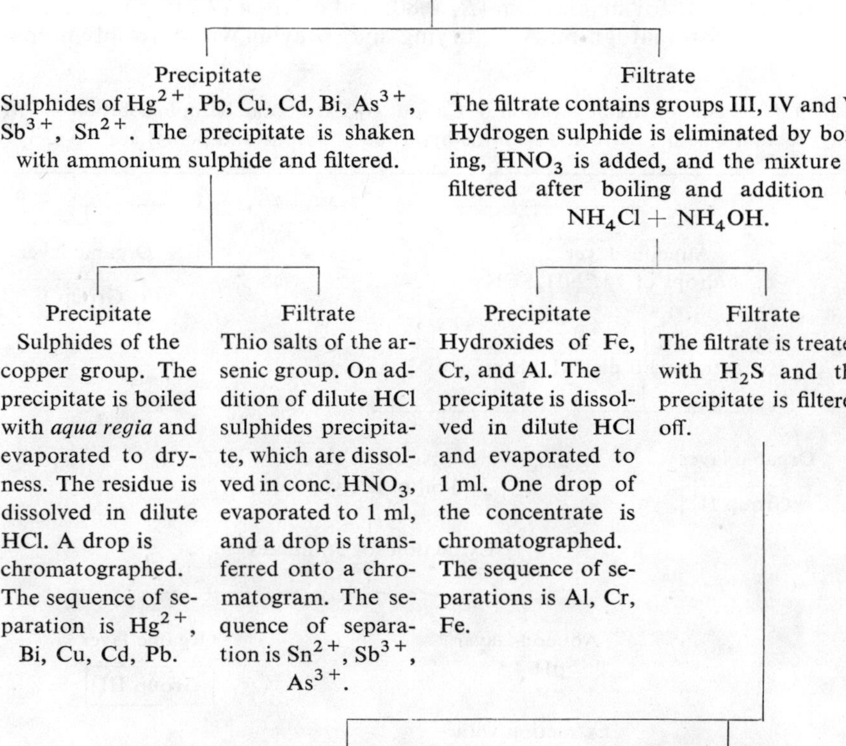

Precipitate
Sulphides of Hg^{2+}, Pb, Cu, Cd, Bi, As^{3+}, Sb^{3+}, Sn^{2+}. The precipitate is shaken with ammonium sulphide and filtered.

Filtrate
The filtrate contains groups III, IV and V. Hydrogen sulphide is eliminated by boiling, HNO_3 is added, and the mixture is filtered after boiling and addition of $NH_4Cl + NH_4OH$.

Precipitate
Sulphides of the copper group. The precipitate is boiled with *aqua regia* and evaporated to dryness. The residue is dissolved in dilute HCl. A drop is chromatographed. The sequence of separation is Hg^{2+}, Bi, Cu, Cd, Pb.

Filtrate
Thio salts of the arsenic group. On addition of dilute HCl sulphides precipitate, which are dissolved in conc. HNO_3, evaporated to 1 ml, and a drop is transferred onto a chromatogram. The sequence of separation is Sn^{2+}, Sb^{3+}, As^{3+}.

Precipitate
Hydroxides of Fe, Cr, and Al. The precipitate is dissolved in dilute HCl and evaporated to 1 ml. One drop of the concentrate is chromatographed. The sequence of separations is Al, Cr, Fe.

Filtrate
The filtrate is treated with H_2S and the precipitate is filtered off.

Precipitate
Sulphides of the zinc group. HCl and $KClO_3$ are added, the mixture is evaporated to dryness and dissolved in water. One drop of the solution is chromatographed. The sequence of separation is Mn, Co, Ni, Zn.

Filtrate
The calcium group and magnesium. It is concentrated and excess of $(NH_4)_2CO_3$ is added. The mixture is filtered after one hour's standing

Precipitate
Contains carbonates of Ca, Sr, and Ba. It is dissolved in dilute HCl, concentrated, and one drop is chromatographed. The sequence of separation is Ca, Ba, Sr.

Filtrate
Magnesium test with 0.1% Magneson solution.

tion decomposes the hexamine, and ammonium salts are eliminated by sublimation at a higher temperature. The residue is wetted with hydrochloric acid, evaporated to dryness, and extracted with water. Alkaline earths are precipitated with ammonium carbonate and ethanol. The washed precipitate is dissolved in hydrochloric acid and chromatographed with a mixture of methanol, hydrochloric acid, and water (80 : 10 : 10). After detection of the dried chromatogram with 8-hydroxyquinoline and ammonia, yellow-green fluorescent spots for magnesium (R_F 0.80) and calcium (R_F 0.55) appear on the still wet chromatogram. After drying and spraying with a freshly prepa-

Table 5.3. Scheme of the Separation of Cations by Extraction. Sample solution in $7M$ hydrochloric acid*, extracted with methyl isobutyl ketone and pentyl acetate (2:1)

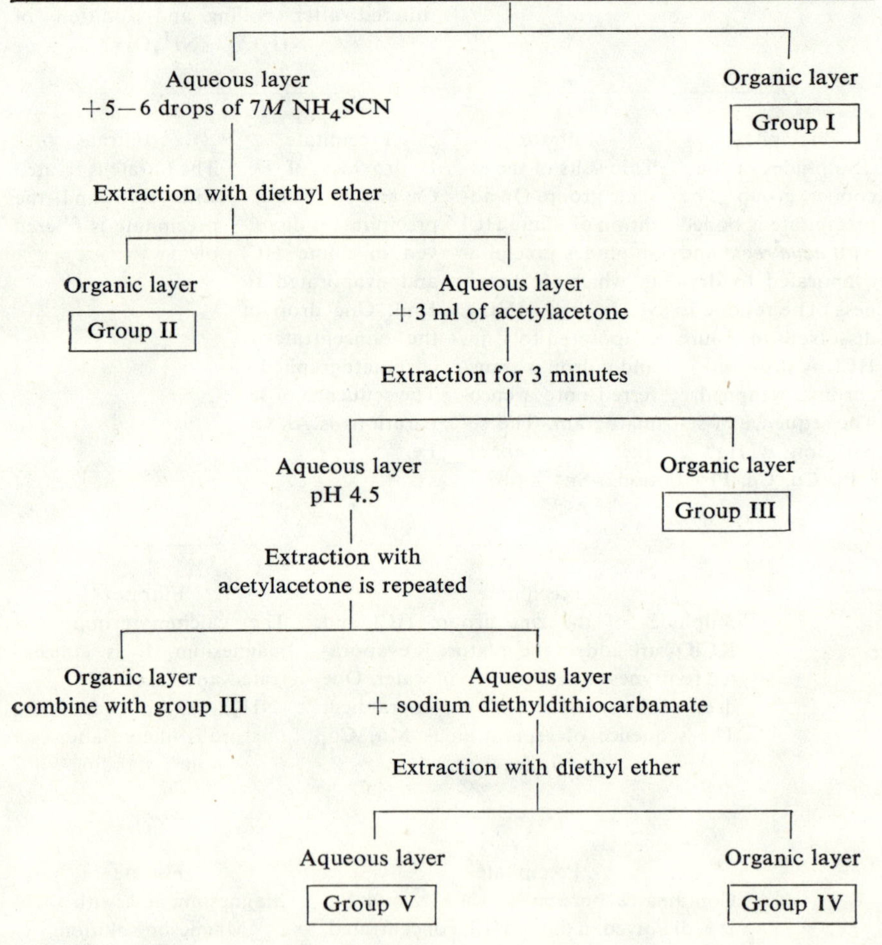

* The chlorides of certain ions, as for example Ag^+, Hg_2^{2+}, Pb^{2+}, W^{6+} and Tl^+, are not included in the system.

Table 5.4. Hashmi's System of Separation of Cations on Thin Layers

Group	Cation	Detection reaction	Solvent R_F		Limit of detection μg	Maximum amount μg	Remarks
I.	Fe^{3+}	a; V	A.	1.00	0.06	24	
	Au^{3+}	a; Gy-Bk		0.71	0.3	15	Light heating after detection
		b; Gy-Bk			0.3	15	
	Mo^{6+}	a; dY		0.62	0.6	67	
	V^{5+}	a; dBl		0.51	0.9	38	Turns lighter after addition of AcOH or heating
	Ga^{3+}	c; RO		0.98	2.0	50	
	Sb^{5+}	c; RO		0.53	2.5	52	Heating after detection
	As^{3+}	c, Y		0.51	6.0	80	Heating after detection
	Te^{4+}	b, dBr		0.51	0.3	10	
	Ge^{4+}	d, R		0.40	0.7	20	
II.	Pd^{2+}	e; YBr	A.	0.95	0.9	50	Appears after a few minutes or after heating
	Pt^{4+}	e; RP		0.90	0.3	45	
	Co^{2+}	e; dBr		0.85	0.2	26	After exposure to ammonia vapour
		f; RV			0.4	26	
	Zn^{2+}	f; RV		0.93	0.3	30	
	Sn^{4+}	f; V		0.67	0.3	30	
III.	Cu^{2+}	g; RBr	B.	0.94	0.15	6	
		e; RBr			0.15	6	In HCl vapour. In ammonia vapour it turns blue
	Ru^{3+}	e; Bl		0.51	0.1	25	
	U^{6+}	g; Br		0.48	1.0	50	
		h; V		1.00	1.0	50	On a silica gel layer
	Ti^{4+}	g; Y		0.39	2.0	40	
	Zr^{4+}	h; V		0.44	1.3	40	
		h; V		0.85	1.3	40	On a silica gel layer
	Cr^{3+}	i; dBr		0.51	1.0	20	

Table 5.4 continued

Group	Cation	Detection reaction	Solvent	R_F	Limit of detection μg	Maximum amount μg	Remarks
	La^{3+}	j; V		0.39	2.0	25	
	Al^{3+}	j; V		0.70	1.0	6	Only on a silica gel layer
IV.	Hg^{2+}	k; Br-Bk	C.	0.91	3.0	15	
		b; Br			3.0	15	
	Cd^{2+}	k; Y		0.82	2.5	50	
	Bi^{3+}	k; dBr		0.56	2.0	120	
	Pb^{2+}	k; Br		0.40	traces		
	Se^{4+}	b; R		0.20	2.0	20	
	Rh^{2+}	b; Y		0.97	0.6	30	
	Ni^{2+}	l; R		0.50	0.3	6	15 to 20 minutes exposure to ammonia vapour
	Mn^{2+}	l; BrY		0.59	4.2	35	
		i; Bl			2.8	35	
V.	on an alumina layer						
	Sr^{2+}	m; R	D.	0.40	2.2	11	
	Ba^{2+}	m; R		0.00	2.8	35	
	Th^{4+}	h; BlV		0.50	1.7	10	
	Ce^{3+}	h; BlV		0.40	2.0	20	
		n; Y			1.3	20	
		o; GBk			2.0	20	
	Mg^{2+}	h; V		0.70	2.5	8	Difficult to detect even after strong heating. NH_4 ion interferes but can be eliminated by previous boiling with NaOH
	K^+	p; Y		0.40	2.0	40	
	Rb^+	p; BkBr		0.33	2.2	45	
	Cs^+	p; OBr		0.24	3.5	40	
	Li^+	q; dBr		0.91	0.4	4	
	on a silica gel layer						
	Sr^{2+}	e; RY	E.	0.88	1.3	11	After heating
	Ba^{2+}	e; R		0.00	2.0	30	After heating
	Mg^{2+}	h; RV		0.89	0.8	4	After exposure to ammonia vapour and heating
	Ca^{2+}	h; BlV		0.80	1.0	5	
	Th^{4+}	h; BlV		1.00	1.3	10	
	Ce^{3+}	h; BlV		1.00	1.5	20	
		n; Y			1.0	20	
		o; BrBk			1.5	20	

red solution of rhodizonic acid (or its sodium salt), strontium and barium appear as distinct red spots with R_F values 0.35 and 0.20, respectively.

The filtrate from the carbonate group is evaporated with hydrochloric acid and ammonium salts are eliminated by sublimation; the residual material is dissolved in hydrochloric acid and chromatographed with methanol, hydrochloric acid, and water (80 : 10 : 10). Alkali metal salts separate in the following order: lithium R_F 0.80, sodium 0.40, and potassium 0.20. They are detected with zinc uranyl acetate and ultraviolet light (Li and Na), or with sodium lead cobaltinitrite (K).

Biswas and Dey's System

Biswas and Dey [10] also separated the mixture of cations into groups by chemical means, but used the same solvent system for all groups, namely 60% aqueous ethanol, with ascending development at 30°, for 90 minutes, on Whatman No. 1 paper. The separated cations were detected with suitable organic reagents. The scheme for this procedure is given in Table 5.2.

Notes to Table 5.4:

Colours: Bk — black, Bl — blue, Br — brown, Gy — grey, O — orange, P — purple, R — red, Y — yellow; d — dark

Reagents:

a — tannin, 10% aqueous solution
b — stannous chloride — potassium iodide; 5 g of $SnCl_2$ is dissolved in 10 ml of conc. HCl, diluted to 100 ml with water and 0.5 g of KI is added
c — dithizone, 2% solution in chloroform
d — phenylfluorone, 0.05% solution in a mixture of 96% ethanol and conc. HCl (3 : 1)
e — rubeanic acid — 0.5% solution in 96% ethyl alcohol
f — diphenylcarbazide, 1% solution in 96% ethanol
g — potassium ferrocyanide, 5% aqueous solution
h — alizarin, saturated solution in 96% ethyl alcohol
i — sodium peroxide — benzidine; the chromatogram is sprayed with 5% aqueous Na_2O_2 and then with 1% benzidine in glacial acetic acid
j — quinalizarin, 0.05% solution in 70% ethanol
k — sodium sulphide, 2% aqueous solution
l — dimethyl glyoxime, 1% solution in 96% ethanol
m — sodium rhodizonate, 1% aqueous solution, freshly prepared
n — hydrogen peroxide–ammonia, a mixture of 30% H_2O_2 and conc. ammonia solution (1 : 1)
o — tannin, 10% solution in a mixture of water and glycerol (1 : 1)
p — three parts of the compound prepared by mixing 11.4 g of $Co(CH_3COO)_2 \cdot 4H_2O$, 16.2 g of $Pb(CH_3COO)_2 \cdot 3H_2O$, 20 g of $NaNO_2$ and 2 ml of glacial acetic acid and dissolving them in water, then diluting to 150 ml, centrifuging and filtering, are mixed with 1 part of methanol
q — silver nitrate–fluorescein; 1% $AgNO_3$ solution. After brief drying the chromatogram is sprayed with 0.1% of alcoholic fluorescein solution

Composition of solvent systems:

A: acetone — hydrochloric acid (4*M*) — acetylacetone (45 : 3 : 2)
B: acetone — hydrochloric acid (4*M*) (47 : 3)
C: acetone — hydrochloric acid (4*M*) — acetylacetone (96 : 3 : 1)
D: acetone — hydrochloric acid (4*M*) (45 : 4)
E: acetone — hydrochloric acid (conc.) (47 : 3)

Separation by Circular Thin-layer Chromatography

Hashmi and co-workers [11] give a procedure for systematic analysis of forty cations and nineteen anions. Theirs is a typical microchemical procedure, as the whole analysis requires only 0.5 ml of sample solution. The separation proper is carried out on a thin layer of alumina with 5% calcium sulphate as binder, or on silica gel without a binder. The authors also devised an application of this system to qualitative analysis of metals.

Before the chromatographic separation and detection of single cations the mixture is first separated by solvent extraction into five groups. The scheme of this extraction is shown in Table 5.3. After application of the sample to the thin layer, chromatography is carried out with one of five proposed solvents followed by chemical detection with some suitable reagent. The scheme of this procedure is given in Table 5.4. Usually alumina layers are employed; silica gel layers are used only in certain cases indicated in the table for the separation of cations of the 5th group.

5.1.2. Differential Detection of Cations
Partially Separated by Chromatography

In these systems of qualitative analysis, cations are not separated chemically into groups, but the chromatographic method is combined with microchemical detection of cations with sensitive organic reagents. The mixture of cations is chromatographed with one or more solvent systems, on several chromatograms which are then sprayed with various reagents. This system was developed by Pollard and co-workers [12, 13, 16] under the name 'Differential Spraying Method'. A partial simplification of this system was proposed by Elbeih [14].

Pollard's System

Pollard's 'Differential Spraying Method' involves running a series of identical chromatograms, in a 1-butanol—benzoylacetone mixture. The cations are identified on various chromatograms by using the following division into groups:

A: Ag, Pb, Hg, As, Sb
B: Cr, Mn, Cu, Co, Ni, Bi, Fe
C: Sn, Cd, Zn, Al, Ca, Ba, Sr, Mg
D: Na, K.

For the analysis, two different solutions of the sample are prepared, (*a*) in 2*M* nitric acid, and (*b*) in 2*M* hydrochloric acid. Some anions can interfere during the chromatographic separation of certain cations, and are therefore as a rule eliminated before chromatography. Tartrate, oxalate, formate, acetate, citrate and the ammonium cation can be eliminated by fusion in sodium carbonate, in which case it is necessary to check beforehand, for example with zinc uranyl acetate and observation under ultraviolet light, whether the sample contains sodium. Phosphate, borate, and some other anions have to be eliminated by passage through a suitable ion-exchange column, as for example of Zeocarb 215, using a $1M$ hydrochloric acid solution of the sample and elution of the retained cations with $5M$ hydrochloric acid.

For the chromatographic separation Whatman No. 1 paper (40×25 cm) is most suitable. Two sheets (I and II) are marked with the start line 8 cm from one end and the sample solutions are applied at 2-cm intervals. On sheet I two spots of solution a are applied and solution b is used for the other spots; on sheet II all the spots are from solution b. Sheet II is then sprayed twice with an alkaline solution of sodium hypobromite, and then with $2M$ acetic acid, being dried under an infrared lamp between sprayings. This precipitates manganese, cobalt and nickel at the start, in the form of insoluble hydrated oxides. Both sheets are then placed in the chromatographic chamber and developed with an acidified mixture of 1-butanol and benzoylacetone (100 ml of 1-butanol + 100 ml of $0.1M$ nitric acid + 10 ml of benzoylacetone). The chromatograms are then dried in a current of warm air and from sheet I both chromatograms of solution a are cut off to be used for the detection of cations of group A. The other chromatograms are sprayed with $2M$ hydrochloric acid and dried at $50-60°$ to destroy all non-ionic complexes. The sheets are then cut into strips which are marked as follows:

Strips from sheet I, solution a, with letter X
Strips from sheet I, solution b, with letter Y
Strips from sheet II, solution b, with letter Z

For the detection of group A, 2 strips X and one Y are taken.
For the detection of group B, 3 strips Y are taken.
For the detection of group C, 4 strips Y and 2 strips Z are taken.
For the detection of group D, 1 strip Y is taken.

It is advantageous to have two sets of such chromatograms, because sometimes repetition may be necessary.

The first strip X is sprayed with potassium chromate ($A1$) and exposed to ammonia ($A2$). This treatment detects Ag, Pb, Hg(I), Cu, and Fe.

The second strip X is sprayed with $2M$ hydrochloric acid, dried, and sprayed again with yellow ammonium sulphide ($A3$). Ag, Pb, Hg, Cd, As and Sb become visible. In order to differentiate between the black sulphides the chromatogram is sprayed with alkaline hydrogen peroxide, fumed with sulphur dioxide and simultaneously warmed over an electric hot-plate at approx. $40°$. Alkaline peroxide oxidizes all black sulphides with the exception of those of Hg(II) and Ni ($A4$).

The first strip Y is sprayed with ammonium sulphide ($A5$), to detect Sb and As. The sulphides of these elements are then dissolved by spraying with concentrated hydrochloric acid and drying over an electric hot-plate. Spraying with a solution of hypophosphite and heating with an infrared lamp at $35°$ (the temperature is checked with a thermometer placed beneath the paper, but kept in contact with it) causes spots to appear for As and Hg. The strip is then sprayed with phosphomolybdic acid ($A6$) and heated, whereupon antimony appears.

Table 5.5. Pollard's System of Qualitative Chromatographic

Chromatogram (R_F)	A1	A2	A3	A4	A5
Reagent	K_2CrO_4	NH_3	1. $2M$ HCl 2. $(NH)_{24}S$	1. Alkali 2. SO_2	1. $(NH_4)_2S$ 2. HCl 3. NaOBr
	Pb, Y	Pb, O	Pb, Br Sb, O Mn, Co, Ni, Br	Ni, R	Mn, Co, Ni, Br
0.1	Ag, R	Ag, Y	Ag, Br Cd, Y		
0.2		Cu, Bl			
0.3					
0.4	Hg_2^{2+}, OBr	Hg_2^{2+}, R	Hg_2^{2+}, R As, Y	Hg_2^{2+}, R	Hg_2^{2+}, R As, Br
0.5	Hg^{2+}, O		Hg^{2+}, R	Hg^{2+}, R	Hg^{2+}, Gy
0.6		Fe, Br			
0.7					

Chromatogram (R_F)	C1	C2	C3a	C3b	C4
Reagent	Oxine NH_3-UV	CH_3COOH UV	1. NH_3 2. Sodium rhodizonate	$2M-$HCl+ $2M-$AcOH (1 : 1)	1. $CuSO_4$ 2. $HgCl_2$ 3. NH_4SCN
	Al, G Ca Sr, Ba Mg, BlG	Al, G	Ba, R Sr	Ba, R	
0.1					
0.2					
0.3					Zn, L
0.4	Zn, Y Cd	Zn, Y			
0.5					
0.6					
0.7	Sn, Y	Sn, Y			

Notes:
Colours: Bl — blue, Br — brown, G — green, Gy — grey, L — lilac, O — orange,
The solvents are described on p. 63.

Analysis of Cations (Differential Spraying Method)

A6	B1	B2	B3	B4	B5	B6
Phospho-molybdic acid	NaOBr	Acid H_2O_2	1. NaOBr 2. SO_2 3. Benzidine 4. NH_3	1. Rubeanic acid 2. NH_3	KSCN-acetone	1. NH_3 2. Dimethyl-glyoxime 3. Rubeanic acid
Sb, M	Mn, Co Ni, Br Cr, Y Cu, Bl Fe, Br	Cr, Bl Cu, Br	Mn, Bl	Co, Ru Ni, Bl Cu, G	Bi, Y Co, Bl Cu, R Fe, RBr	(Co) Ni, R (Cu)

C5	C6	C7	D1	D2
Quinalizarin	Alkaline H_2O_2	Gallaceto-phenone	Zinc uranyl acetate-UV	Lead cobalti-nitrate
Al, V Zn, VR	Mg, Bl	Ba, Sr, Ca, P	Na, BlG	K, GyR

P — purple, R — red, Ru — rusty, V — violet, Y — yellow

Another strip Y is sprayed with sodium hypobromite and heated ($B1$). This detects Fe, Pb, Cu, Mn, Co, Ni, Cr. The strip is then sprayed with hydrogen peroxide acidified with sulphuric acid ($B2$); Mn, Co, Ni and Cr become visible.

Another strip Y is sprayed first with sodium hypobromite, and is then exposed to a stream of sulphur dioxide in order to reduce chromate to chromium (III) and to decompose the excess of hypobromite ($B3$). The manganese spot is sprayed with a small excess of benzidine solution, exposed to ammonia and resprayed with benzidine. After drying, the chromatogram is sprayed with rubeanic acid and fumed with ammonia ($B4$) to detect Cu, Co and Ni.

Another strip Y is sprayed with a mixture of equal parts of a saturated thiocyanate solution and acetone ($B5$); Bi, Fe, and Co appear. On exposure to ammonia ($B6$) the spots disappear. The chromatogram is sprayed with a solution of dimethylglyoxime (for Ni) and finally with a solution of rubeanic acid (for Cu and Co).

Two strips Y and two strips Z are sprayed with 8-hydroxyquinoline ($C1$), exposed to ammonia and observed under ultraviolet light for fluorescent spots. They are then sprayed lightly with glacial acetic acid ($C2$). Some fluorescent zones (Ca, Sr, Be, Mg, Cd) disappear, others such as Al, Sn and Zn remain visible.

Chromatogram $C2$ is exposed to ammonia and sprayed with rhodizonate ($C3$a). Spots of strontium and barium become visible. They can be differentiated by spraying the chromatogram with a mixture of equal parts of $2M$ hydrochloric acid and $2M$ acetic acid ($C3$b). If the red spot slowly vanishes it is due to Sr; if it becomes brighter red and permanent, it is due to Ba.

If cobalt is absent zinc can be detected by spraying the chromatogram Y with copper sulphate solution and then with a solution of 2.7 g of mercuric chloride and 3 g of ammonium thiocyanate in 100 ml of water ($C4$). If cobalt is present the procedure consists of spraying first with a saturated potassium nitrite solution in $2M$ acetic acid, thus transforming cobalt into the yellow potassium cobaltinitrite, and then with cupric sulphate and a solution of mercuric chloride and ammonium thiocyanate.

Spraying the chromatogram Y with quinalizarin ($C5$) brings about the appearance of aluminium and zinc. Subsequent spraying with alkaline hydrogen peroxide causes the spots to disappear ($C6$), but at the same time the spot of magnesium appears.

Calcium can be detected on a special chromatogram Y by spraying with a solution of galloacetophenone or pyrogallol carboxylic acid in ammonia ($C7$).

For the detection of sodium, chromatogram Y is sprayed with $2M$ acetic acid, dried, and sprayed again with zinc uranyl acetate ($D1$). On further spraying of this chromatogram with lead cobaltinitrite and finally with saturated sodium nitrite solution ($D2$) potassium appears as a grey-black spot. The sulphate ion interferes with this detection; if present it should be first precipitated as barium sulphate by spraying the chromatogram with a barium nitrate solution.

A summary of the whole process is given in Table 5.5. This system has been further extended to the detection of cerium, lithium, molybdenum, thallium, titanium, uranium, vanadium and tungsten [16].

Elbeih and Gabra's System

Elbeih and Gabra [14] proposed a procedure for the detection of thirty cations, based on separation of the mixture by means of a solvent system containing mainly dioxan and EDTA, and on detection with a mixture f kojic acid and 8-hydroxyquinoline in addition to specific reagents. They used chromatography on circular chromatograms 27 cm in diameter; after development the chromatogram was cut into 29 radial segments for detection.

The preparation of the sample, as well as the elimination of the interfering anions is the same as in Pollard's system (see preceding section). For specific detection of aluminium, aluminon (ammonium salt of aurinetricarboxylic acid) is used, ammonium sulphide for antimony, silver nitrate for arsenic, sodium rhodizonate for barium, strontium and lead, p-nitrobenzeneazore-

Table 5.6. The System of Qualitative Analysis of Cations
According to Elbeih and Gabra

| Ion | Colour | | R_F | Ion | Colour | | R_F |
	kojic acid-oxine UV light	$(NH_4)_2S$			kojic acid-oxine UV light	$(NH_4)_2S$	
Sb^{3+}	—	O	0.00	Pb^{2+}	Gy	Br	0.17
Ba^{2+}	WBl	—	0.02	V^{5+}	Bk	Br	0.17
Cr^{3+}	V	1G	0.04	Fe^{3+}	Bk	Bk	0.23
K^+	V	—	0.04	Cd^{2+}	Y	Y	0.24
Na^+	V	—	0.06	Be^{2+}	W	—	0.26
Sr^{2+}	WBl	—	0.06	Li^+	V	—	0.28
Ag^+	Y	Br	0.08	Cu^{2+}	Bk	Br	0.29
Ti^{4+}	Bk	Y	0.11	Mo^{6+}	Br	—	0.42
Co^{2+}	Bk	Bk	0.12	As^{3+}	—	Y	0.43
Ni^{2+}	Bk	Bk	0.12	Ce^{3+}	V	1G	0.45
Zn^{2+}	Y	—	0.12	Hg^{2+}	Br	Br	0.55
Mg^{2+}	W	—	0.13	Bi^{3+}	Br	Bk	0.69
Mn^{2+}	Bk	1Br	0.15	Th^{4+}	YG	Y	0.73
Al^{3+}	WY	—	0.16	U^{6+}	Bk	O	0.75
Ca^{2+}	WBl	—	0.17	Sn^{2+}	Y	Y	0.82

Notes: The solvent system used had the following composition: 2 ml of water were mixed with 2 ml of conc. nitric acid and 0.05 g of EDTA were added and completely dissolved by heating. After cooling, 100 ml of dioxan were added followed by 1 g of antipyrine, which was dissolved in the cold. The mixture was stable for 3 days, after this it turned yellow.

Detection mixture: 0.5 g of kojic acid and 2.5 g of 8-hydroxyquinoline are dissolved in 500 ml of 60% ethanol. Ammonium sulphide is applied to the dried chromatogram after detection with this mixture.

Colours: Bk — black, Bl — blue, Br — brown, G — green, Gy — grey, O — orange, V — violet, W — white, Y — yellow, 1 — light.

sorcinol for beryllium and magnesium, cinchonine and potassium iodide for bismuth, Titan Yellow for cadmium, Murexide (after preliminary spraying with potassium cyanide solution) for calcium, benzidine for cerium and manganese, diphenylcarbazide for chromium, 1-nitroso-2-naphthol for cobalt, diethyldithiocarbamate for copper, ferrocyanide for iron and uranium, diphenylcarbazone for mercury and thorium, potassium thiocyanate and acetone for molybdenum, dimethylglyoxime for nickel, dipicrylamine — fluorescein for potassium, p-dimethylaminobenzylidenerhodanine for silver, zinc uranyl acetate for sodium and lithium, dithiol for tin, chromotropic acid for titanium, tannin for vanadium, and 8-hydroxyquinoline for zinc.

The corresponding R_F values and colours given by cations after spraying with a mixture of kojic acid and 8-hydroxyquinoline followed by ammonium sulphide are listed in Table 5.6.

5.1.3. Chromatographic Detection of Cations

The third group of systems for the detection of cations in an unknown mixture consists of chromatography and identification of the cation on the basis of its R_F values in several solvent systems. The system by Ritchie [15] may serve as a typical example; Pollard's comparative method [13, 16] extended by Elbeih [17] also deserves attention.

Ritchie's System

Ritchie [7] proposed the term 'chromatographic profile' for the result of a method based on the determination of R_F values. The following four solvent systems were used:

A: 1-butanol, concentrated hydrochloric acid, concentrated hydrofluoric acid and water (1000 : 50 : 2 : 48).

B: the alcoholic layer of a mixture of 1-butanol, concentrated hydrobromic acid and water (100 : 10 : 90).

C: ethanol, methanol and hydrochloric acid (45 : 45 : 60).

D: acetone, concentrated hydrochloric acid, concentrated hydrofluoric acid and water (90 : 5 : 1 : 4).

For the detection of the majority of cations yellow ammonium sulphide and 8-hydroxyquinoline solution suffice. The use of selective reagents should be kept to a minimum. However, it is advantageous to develop at least four chromatograms in every solvent mixture and detect the first with yellow ammonium sulphide and stannous chloride (possibly with addition of potassium thiocyanate), the second with 8-hydroxyquinoline at pH 2−6, the third with the same reagent but at pH 6−10 and subsequent treatment with sodium rhodizonate, the fourth with chlorine water and p-dipicrylamine. In Table 5.7 the corresponding R_F values and colours after the detection with 8-hydroxyquinoline are given.

Pollard's Comparative System

In this system [13, 16] chromatograms are prepared in three different solvent systems and the R_F values obtained are compared with previously determined R_F values of known cations in the same systems (Table 5.8). As chromatographic solvents 1-butanol—benzoylacetone, collidine, and

Table 5.7. Ritchie's System of Qualitative Chromatographic Analysis of Cations
(Chromatographic Profiles)

Ion	Colour with $(NH_4)_2S$	Colour with 8-hydroxyquino-line		R_F values			
		Visible light	UV light	Solvent A	Solvent B	Solvent C	Solvent D
Ag^+	Bk	Y	fY	0.77C	1.0C	0.13C	0.59C
Pb^{2+}	BkBr	Y	fY	0.44C	0.64C	0.45C	0.48C
Cd^{2+}	Y	GY	fY	0.92	1.0	0.96	0.77–0.83
Bi^{3+}	Br	Y	fY	0.75	0.85	0.88	0.85
Sn^{2+}	BrBk	Y	fY —	0.93	0.95–1.0	0.93	0.82
Hg^{2+}	Bk	Y	fG	1.0	1.0	1.0	1.0
Sb^{3+}	Y	YBr		0.88	1.0	0.80C	0.82
As^{3+}	Bk	YBk		0.92	0.92	0.92	0.65
Fe^{3+}	Bk	Bk		0.79–0.93	0.65–0.75	0.70–0.73	0.86
Cu^{2+}	Br	YBr		0.53	0.54; 1.0	0.65	0.56
Co^{2+}	BrBk	BrY		0.45	0.30–0.38	0.70	0.45–0.55
MoO_4^{2-}	Y	Y		0.75	0.53C	0.60	0.83–0.92C
Tl^+	Br	YBr		0.41C	1.0C	0.41C	1.0C
Tl^{3+}	Br	Y		1.0	1.0C	1.0C	1.0
Ni^{2+}	BrBk	GBr		0.52	0.32	0.63–0.67	0.03
Mn^{2+}	BrBk	BrBk		0.48	0.32	0.70	0.25
Sc^{3+}		Y	fYG+	0	0.25	0.75	0.0
Y^{3+}		Y	fY	0	0.18	0.24–0.70	0.0
La^{3+}		Y	fGY+	0–0.10C	0.19	0.65	0.24C
Ga^{3+}		Y	fY +	1.0	0.65–0.70	0.80	0.63–0.68
In^{3+}		Y	fY +	0.57	0.91–1.0	0.78	0.85–0.90
Zr^{4+}		Y	fY +	0.52–0.60C	0.12C	0.52C	0.48C
Ge^{4+}		Y	fY +	0.59	0.81	0.47	0.29
Ca^{2+}		Y	fY +	0.40C	0.19	0.44	0.02
Sr^{2+}			fY +	0.30	0.13	0.40	0.0
Ba^{2+}			fGY	0.22	0.06	0.24	0.0
Zn^{2+}		GY	fGY	0.94	1.0	0.94	0.90
Al^{3+}			fYG+	0.42–0.46	0.36	0.75	0.02
Li^+			fBl +	0.46	0.40	0.75	0.13
Rb^+			fY —	0.43	0.23	0.35	0.05

Table 5.7 continued

Ion	Colour with 8-hydroxyquinoline		R_F values			
	Visible light	UV light	Solvent A	Solvent B	Solvent C	Solvent D
Cs^+		fY	0.46	0.26	0.34	0.03
K^+		fBl	0.41	0.10	0.46	0.05
Nb^{5+}	Y	fY +	0.88	0.26	0.86	0.82–0.94
Ta^{5+}		fBl –	1.0	0.90C	1.0C	1.0
Be^{2+}	Y	fY	0.54–0.60	0.60	0.86	0.22
Mg^{2+}	Y	fY +	0.47	0.37	1.0	0.03
Hf^{4+}	Y	fY +	0.53C	0.11C	0.50C	0.46
Th^{4+}	Y	fO +	0.0	0.15	0.60	0.0
Au^{3+}	Bk		1.0	0.86–1.0	0.91	0.97
Pt^{4+}	Br		1.0C	1.0C	0.90C	0.93C
Pd^{2+}	Y		0.83	1.0	0.91	0.90
Ru^{3+}	Bk		0.30C	1.0C	0.70C	0.82C
Cr^{3+}	G +		0.38	0.23	0.73	0.06
CrO_4^{2-}	Y		1.0C	0.30C	0.86C	1.0C
Ce^{3+}	Br		0.0	0.15	0.54	0.0
Ti^{4+}	YBr –		0.64	0.43	0.76	0.64
UO_2^{2+}	Br		0.46	0.37–0.43	0.76	0.58–0.64
V^{5+}	Br +		0.47	0.43	0.74	0.64
WO_4^{2-}	Br	fY	0.70	0.0	1.0C	1.0C
Ir^{4+}			0.47	0.17	0.82–0.86	0.03
TeO_3^{2-}			0.86	1.0	0.62	0.71; 0.15
TeO_4^-			0.68C	1.0C	0.62C	0.71; 0.15C
SeO_3^{2-}			0.72	1.0	0.72	1.0
ReO_4^-			0.82	1.0	0.87	0.89
Rh^{3+}			0.64	0.43	0.65	0.05

Abbreviations used: Bl — blue, Bk — black, Br — brown, G — green, O — orange, Y — yellow; f — fluorescence, C — comet — shaped spot, + — intensified with ammonia, — — weakened with ammonia
The solvents are described on p. 68.

dioxan — antipyrine are employed. For general detection ammonium sulphide and an alcoholic 8-hydroxyquinoline solution may be used. In addition, rubeanic acid can be used for the detection of copper, cobalt, and nickel, rhodizonic acid for barium, strontium, lead and thallium, violuric acid for lithium, zinc, magnesium, copper, calcium, sodium, potassium and iron, and dithizone for lead, zinc, copper, mercury, silver, cobalt, cadmium and bismuth.

Table 5.8. Comparative System of Qualitative Chromatographic Analysis of Cations According to Pollard

R_F	Solvent		
	A	B	C
	Sb	Pb, Bi, Al, U, Fe, Th	Sb
	Li		Cr
	Bi, Ba	Be, Ti, Hg_2^{2+}, Hg^{2+}	Ba
	Pb, Al, Cr, Ni		K, Al, Li
	Sr, Ce	Cr, Sn^{2+}, Sn^{4+}	Na, Mg, Sr
	Zn, Cd, K, Ca		Co, Ni, Ti
	Co, Mg, Na	Ce	Ag, Zn, Be
	Mn		Mn
0.1	Ag		Fe, Ca
		V	Pb
0.2	Th	Mo	Cd, As
	Cu		
	Hg_2^{2+}	Ba	Cu
	Ti		
			Ce
0.3	Hg^{2+}	K	
	Be		Mo
	U	Sb	
0.4	As	Sr	Hg^{2+}
		Na	
			Hg_2^{2+}, V
0.5	Sn^{4+}, Mo	Ca	
		Li	
	Sn^{2+}		
0.6			Bi
		Mg, As	Th
			Sn^{4+}
0.7		Mn	U
		Co	
		Zn	
		Cu, Cd, Ni	
		Ag	
0.8			
0.9		Fe	

5.2. Systems of Qualitative Analysis of Anions

The systems for qualitative analysis of anions are less well developed than those for cations. When analysing a mixture by paper chromatography two procedures may be used. Depending on which anions are expected to be present, suitable solvents may be chosen from Table 6.5 (e.g. 12, 13, 14) and the observed R_F values compared with those in a table of R_F values. The presence of anions can be confirmed by chromatography with a suitable solvent system and by using selective detection methods if any are available.

Elbeih and Gabra [14] proposed the method of differential detection of chromatographically partially separated anions. They used a mixture of ethanol, pyridine, ammonia and water (60 : 20 : 4 : 16), and circular chromatography. As a general, non-specific, reagent they proposed a solution of 0.05 g of fluorescein in a mixture of 50 ml of 0.4M ammoniacal silver nitrate solution and 50 ml of saturated aqueous 5-aminonaphthalene-1-sulphonic acid solution. Table 5.9 gives corresponding colours and R_F values.

A similar procedure is applied in thin-layer chromatography. The anions are classified into four groups on the basis of their reaction with five different detection reagents [11], namely silver nitrate, potassium iodide, hydrochloric acid, ferric chloride, and potassium dichromate. In cases where the reaction used cannot identify an ion, an additional detection reaction is necessary. Chromatography is usually carried out on thin layers of alumina. Only the anions of group IV are separated on silica gel thin layers. As solvent, a mix-

Table 5.9. System of Qualitative Chromatographic Analysis of Anions
According to Elbeih and Gabra

| Anion | Colour | | R_F value | Anion | Colour | | R_F value |
	visible light	UV light			visible light	UV light	
F^-	1Gy	Gy	0.01	CO_3^{2-}	GY	GY	0.10
$Fe(CN)_6^{4-}$	Y	Y	0.01	IO_3^-	Y	Y	0.10
AsO_4^{3-}	GY	GGy	0.03	AsO_2^-	Y	GGy	0.12
CrO_4^{2-}	Br	Gy	0.03	SO_3^{2-}	Br	Gy	0.20
$Cr_2O_7^{2-}$	Br	GY	0.03	BrO_3^-	Y	Y	0.41
PO_4^{3-}	YG	YGy	0.03	NO_2^-	Y	Y	0.43
S^{2-}	Gy	GGy	0.03	Cl^-	Gy	Y	0.45
CN^-	YG	YG	0.06	Br^-	Gy	Gy	0.50
SO_4^{2-}	Y	Y	0.07	NO_3^-	Y	Y	0.56
$Fe(CN)_6^{3-}$	Pi	RV	0.08	ClO_3^-	Y	Y	0.60
$B_4O_7^{2-}$	Gy	Bl	0.09	I^-	R	V	0.61
$S_2O_3^{2-}$	Br	VGy	0.09	SCN^-	Pi	BrY	0.66

Colours: Bl — blue, Br — brown, G — green, Gy — grey, Pi — pink, R — red, V — violet, Y — yellow; l — light.

Table 5.10. Hashmi's System of Separation of Anions on Thin Layers

Group	Anion	Detection and colour	R_F	Limit of sensitivity μg	Maximum separable amount μg	Remarks
I.	AsO_2^-	a; Y	0.00	0.2	5.0	
	AsO_4^{3-}	a; O	0.00	0.5	7.0	
	$H_2PO_4^-$	a; G	0.00	0.3	2.0	
		b; Bl		1.0	3.0	
	CrO_4^{2-}	a; YO	0.10	0.1	3.0	
	$S_2O_3^{2-}$	a; Bk	0.25	0.5	7.5	
	Cl^-	a; Bl	0.52	1.0	8.0	
	Br^-	a; Br	0.63	0.7	17.0	
	I^-	a; Br	0.76	0.5	20.0	
		b; Br		0.9	21.0	
		c; BlG		1.5	20.0	
	S^{2-}	a; Bk	0.50	0.3	5.0	
		c; RV		2.5	8.0	
II.	IO_4^-	d; Br	0.00	1.0	8.0	
	IO_3^-	d; Y	0.27	0.2	8.0	
	BrO_3^-	d; Y	0.27	0.2	8.0	
	NO_2^-	d; YBr	0.62	0.2	3.0	
		a; Bk		0.7	4.0	after prolonged heating
		e; Br		1.0	4.0	
	ClO_3^-	d; Y	0.65	2.7	6.0	
III.	$Fe(CN)_6^{3-}$	b; G	0.37	0.3	9.0	
	$Fe(CN)_6^{4-}$	b; Bl	0.00	0.2	10.0	
	CH_3COO^-	b; Br	0.10	2.0	13.0	
	SCN^-	b; Br	0.74	0.2	3.5	
IV.	SO_3^{2-}	f; Br	0.21	2.5	20.0	on a silica gel layer
	AsO_2^-	f; G	0.65	9.8	4.0	on a silica gel layer
		a; Y		0.2	5.0	on a silica gel layer

Notes:

Colours: Bk — black, Bl — blue, Br — brown, G — green, O — orange, R — red, V — violet, Y — yellow

Detection reagents:
a — silver nitrate, saturated solution
b — ferric chloride, 10% solution containing $2M$ HCl
c — sodium nitroprusside, 20% solution
d — potassium iodide — HCl, 10% solution in 10 ml of $2M$ HCl
e — ferrous sulphate, 10% solution containing 20 ml g $2M$ H_2SO_4
f — potassium dichromate, saturated solution containing $2M$ H_2SO_4

Solvent system: 1 — butanol–pyridine–water–ammonia (8 : 4 : 8 : 1). The mixture is shaken and the upper layer is used for chromatography

ture of 1-butanol, pyridine, water and ammonia (8 : 4 : 8 : 1) is used. A scheme is shown in Table 5.10.

Kawanabe and co-workers [18] also used thin-layer chromatography for the separation of anions. They separated thiocyanate, iodide, bromide, chloride, ferricyanide, ferrocyanide, chlorate, bromate, iodate, and nitrate in a mixture of acetone and water (10 : 1). R_F values decrease in the sequence given above. A mixture of methanol, 1-butanol and water (3 : 1 : 1) is suitable for the separation of fluoride, nitrite, thiosulphate, sulphate, chromate, phosphate, arsenate and arsenite.

REFERENCES

1. SCHNEER-ERDEY, A. *Talanta* **10**, 591 (1963).
2. ELBEIH, I. I. M. and ABOU-ELNAGA, M. A., *Anal. Chim. Acta* **23**, 30 (1960).
3. ĆELAP, M. B. and RADIVOJEVIĆ, Z. M., *Ber. Chem. Ges. Belgrad* **23/24**, 179 (1958/1959).
4. SEN, B. N. and RAY, K. *J. Sci. Ind. Res.* (*New Delhi*) *Sect. B* **14**, 421, 604 (1955).
5. TAMURA, Z. and ASHIKAWA, R., *Bunseki Kagaku* **3**, 475 (1954).
6. TAMURA, Z. and MIYAZAKI, M., *Bunseki Kagaku* **5**, 566 (1956).
7. TAKITANI, S., FUKAZAWA, M., HASEGAWA, H., *Bunseki Kagaku* **12**, 1156 (1963).
8. PFEIL, E. *Österr. Chem.-Ztg.* **65**, 177 (1964).
9. BLASIUS, E. and GÖTTLING, W., *Z. Anal. Chem.* **162**, 423, (1958).
10. BISWAS, S. D. and DEY, A. K., *Z. Anal. Chem.* **192**, 376 (1963).
11. HASHMI, M. H., SHAHID, M. A., AYAZ, A. A., CHUGTAI, F. R., HASSAN, N. and ADIL, A. S., *Anal. Chem.* **38**, 1554 (1966).
12. POLLARD, F. H., McOMIE, J. E. W. and ELBEIH, I. I. M., *J. Chem. Soc.* 470 (1951).
13. POLLARD, F. H., McOMIE, J. F. W. and STEVENS, H. M., *J. Chem. Soc.* 771 (1951).
14. ELBEIH, I. I. M. and GABRA, G. G., *Chemist-Analyst* **52**, 36 (1963).
15. RITCHIE, A. S., *J. Chem. Educ.* **38**, 400 (1961).
16. POLLARD, F. H., McOMIE, J. F. W. and STEVENS, H. M., *J. Chem. Soc.* 1863 (1951).
17. ELBEIH, I. I. M. and ABOU-ELNAGA, M. A., *Anal. Chim. Acta* **17**, 397 (1957).
18. KAWANABE, K., TAKITANI, S., MIYAZAKI, M. and TAMURA, Z.: *Bunseki Kagaku* **13**, 976 (1964).

Chapter 6

A REVIEW OF CHROMATOGRAPHIC SEPARATION OF INORGANIC COMPOUNDS

Chromatographic methods of separation, with the exception of ion-exchange chromatography, are used less in inorganic than in organic analysis. In the analysis of complex organic mixtures it often happens that a suitable method for a reliable and sensitive determination of a given component is available, but accompanying products make its determination difficult or impossible. In that case an analytical procedure should be chosen which would eliminate the disturbing effect of accompanying components. Such a method is chromatography, which if applied in a suitable form gives sharp and quantitative separation.

Chromatographic methods can also be successfully used for the qualitative and semiquantitative analysis of complex inorganic mixtures. The scope of chromatography should obviously be known in order for a suitable chromatographic method to be selected. Therefore, in subsequent chapters known separation methods involving chromatographic analysis will be reviewed. Both adsorption and partition methods will be discussed. Partition chromatography, especially paper and thin-layer chromatography, is the most important in the separation of inorganic mixtures. For rapid orientation, those solvent systems are tabulated which permit separation of one ion from others, and R_F values are listed, characterizing the suitability and sharpness of the separation. Only the more important of the many published methods are included in the text or tables.

6.1. Adsorption Chromatography

6.1.1. Alumina Columns

The first investigations in the field of inorganic adsorption analysis were published by Schwab and his school [1 – 6] and he is mainly responsible for its initial development. He studied the sorption of cations on a column of alumina by chromatographing systematically a pair of cations in a non-complexing medium and determined the sequence in which they were sorbed on the chromatographic column. His investigations resulted in the following sequence, the cations on the left displacing those on the right:

$$\begin{array}{ll} & \text{Co}^{2+} \\ \text{Cr}^{3+} & \text{Ni}^{2+} \\ \text{H}^+,\ \text{Fe}^{3+},\ \text{UO}_2^{2+},\ \text{Pb}^{2+},\ \text{Al}^{3+},\ \text{Cu}^{2+},\ \text{Ag}^+,\ \text{Zn}^{2+},\ \text{Cd}^{2+},\ \text{Tl}^+,\ \text{Mn}^{2+} \\ \text{Hg}^{2+} & \text{Fe}^{2+} \end{array}$$

However, a sharp separation is obtained only if a small number of elements is chromatographed. With complex mixtures, separation is often not clear, mainly because the zones of separated elements in inorganic adsorption chromatography immediately follow each other. Schwab endeavoured to explain the principle of the retention of ions on alumina, and came to the conclusion that the most important factor is the exchange of ions. Most aluminas contain sodium in the form of carbonate or aluminate. According to Schwab this sodium may take part in the exchange reaction:

$$n(>\text{Al—O—Na}) + \text{M}^{n+} \rightleftharpoons (>\text{Al—O})_n\text{M} + n\text{Na}^+$$

Both Venturello [7], who also worked on the separation of cations, and Flood [8], who separated cations on alumina-impregnated paper, confirmed Schwab's adsorption sequence and also considered ion-exchange as a possible principle of separation.

The separation of inorganic ions on alumina can also be considered from two other viewpoints; physical adsorption of ions and molecules, and hydrolysis and precipitation. Jacobs and Tompkins [9] believe that the sorption of cations when moving with anions depends on molecular adsorption. The strength of the equivalent adsorption of cations and anions is proportional to the covalent character of the salt. Neugebauer and Schäfer [10] also propose a primary effect of physical adsorption. They carried out many successful separations on a column of especially pure alumina, free from alkali metals and alkaline earth metals. They eluted with pure water and determined the ions by conductimetry.

In explaining the separation of inorganic ions on alumina many authors incline to the opinion that hydrolytic and precipitation processes also have a substantial effect on separation. Schwab himself [11] stated in a later paper, basing his opinion on absorption spectra, that during separation basic double salts are formed, whereas Stewart [12] proposed the precipitation of carbonates. These theories were modified and made more accurate by many other investigations. The explanation of the retention of ions on alumina in these terms is not easy, however, and in addition to this they do not explain the separation satisfactorily. In fact, probably all the factors mentioned play a part, and the separation of inorganic ions on alumina cannot yet be fully explained. This is also true of other adsorbents.

Anions can be separated on alumina in the same way as cations. Schwab [3], in systematic experiments on alumina washed with acid, deduced the following adsorption sequence of anions:

$$\text{OH}^-,\ \text{PO}_4^{3-},\ \text{F}^-,\ \begin{array}{c}\text{Fe(CN)}_6^{4-}\\ \text{CrO}_4^{2-}\end{array},\ \text{SO}_4^{2-},\ \begin{array}{c}\text{Fe(CN)}_6^{3-}\\ \text{Cr}_2\text{O}_7^{2-}\end{array},\ \text{Cl}^-,\ \text{NO}_3^-,\ \text{MnO}_4^-$$

Table 6.1. Examples of the Use of Adsorption Chromatography of Inorganic Compounds on Alumina

Ions separated	Remarks	Reference
As, Sb, Sn	—	[13]
As, Sb, Sn, Bi	For the separation of tin	[14]
As, Sb, Sn	—	[15]
Pb	On a column of alumina containing potassium dichromate. Quantitative evaluation possible from the width of the zone (3% error)	[16]
Fe, Al, Cr, Zn, Mn, Ni, Co	Next to each other. For a qualitative analysis	[17]
Fe, Cu, Zn, Mn	In ash from tomatoes	[18]
Co, Ni, Cu	pH 7.6. Sensitivity down to 0.001 µg of Co and Ni, and 0.05 µg of Cu	[19]
Fe	Detection down to 0.02% Fe	[20]
Zn, Cd, Mn, Mg, Ca Sr, Ba	Purification of the salts of these elements from trace impurities of Fe, Co, Ag, Pb	[21]
Cd	Elimination of Fe and Cu down to 3×10^{-5} or 1×10^{-6}%	[22]
Zn	On a column of alumina and ZnO — for the elimination of Fe and Cu, down to 3×10^{-5} or 1×10^{-6}%	
Fe, Cu, Co	—	[23]
Fe, Cu, Co	Separation of Fe	[24]
Co	Colorimetric determination after separation	[25]
Cu, Co, Ni, Fe, Cr	Determination of Co in steel, and Cu and Ni in alloys and cast iron	[26]
Be, Mg	Detection with quinalizarin. Semiquantitative determination	[27]
Pt-metals	Sequence of adsorption Ir-Pt-Pd-Rh. Evaluation by X-rays	[28]
Rare earths	Elution monitored by spectrometry. Adsorption strongest for Y	[29]
	Radiometric monitoring and quantitative determination	[30]
	Separation of La and Ce	[31]
Cl^-, Br^-, I^-	Radiometric evaluation	[32]
Cl^-, Br^-, I^-	Separation on columns, thin layers or paper coated with aluminium oxide. Determination by spot tests, radiometrically or polarographically	[688]
SO_4^{2-}	Separation of H_2SO_4, for the determination of S in steel	[33]

In view of the position of the chloride and nitrate ions in this series, hydrochloric or nitric acid can be used for the preparation of separation columns.

Among numerous practical applications, adsorption chromatography on alumina has been used to eliminate trace impurities from solutions of pure salts. Some other examples are given in Table 6.1.

The complexation of inorganic ions with organic reagents is sometimes used with advantage; in such cases the separation loses the features of inorganic adsorption chromatography and becomes similar to the separations usual in organic analyses. Erämetsä [29a] proposed the separation of dithizonates of antimony, tin, nickel, manganese, copper, cadmium, iron(II), cobalt, zinc, and mercury(II) with carbon tetrachloride on weakly active alumina. If the separation is carried out in chloroform the complexes are adsorbed much more feebly, so that alumina prepared according to Brockmann may be used.

6.1.2. Other Inorganic Adsorbents

Schwab [2] used other inorganic adsorbents, in addition to alumina, for chromatographic separations. He investigated various oxides, glass powder, and barium sulphate, and found them for the most part inactive in the separation of cations. Zinc oxide and magnesium oxide, however, he considered too active. He indicated silica gel and calcium phosphate as satisfactory, but did not develop their use or indicate the sequence of adsorption of ions. Various other authors later described in greater detail some adsorbents on which a sharp separation of various cations was possible. Bach [34] used zinc sulphide for the separation of copper and cadmium, Gapon [35] used barium aluminate and zinc oxide for the separation of various cations, and Kraus [36 – 38] used zirconium oxide and zirconium phosphate. Sen used chalk to separate the following pairs of cations: Cu—Sb, Bi—Sb, Cd—Sb, Cd—Sn, Hg—Sb, Cu—As, Hg—Cu, Cu—Fe [39] and also made a qualitative analysis of the copper group [40] with chalk as adsorbent. In similar manner he was able to separate pairs of anions: PO_4^{3-}—HPO_3^{2-}, AsO_4^{3-}—AsO_3^{3-}, AsO_4^{3-}—HPO_3^{2-}, AsO_3^{3-}—HPO_3^{2-}, $S_2O_3^{2-}$—SCN^-, $Fe(CN)_6^{4-}$—SCN^-, $Fe(CN)_6^{3-}$—SCN^-, $Fe(CN)_6^{4-}$—$Fe(CN)_6^{3-}$ [41, 42]. Glemser and Rieck [43] described the utilization of ferric oxide, which can be prepared with various degrees of activity and which they claim is much more suitable than alumina. However, they used it only for separations in organic analysis. For the separation of copper and its determination in steel and various alloys, cadmium sulphide was found just as convenient [44].

Numerous authors have shown that silica gel may be used in adsorption chromatographic analysis. On a silica gel column zirconium and hafnium have been separated [45] as have the rare earths [46, 47] and iron has been determined in concentrated sulphuric acid [48]. Silica gel is also a suitable adsorbent for the retention of some organic metal complexes. The 1,10-phenanthroline iron(II) complex has been adsorbed and determined [49, 50], as have a number of metal-dithizone [51], and metal-oxine complexes [52, 53]. Šulcek and co-workers separated small amounts of uranium [54, 55] and beryllium [56] on silica gel, using EDTA for separation. They also developed procedures for the determination of these elements in minerals and mineral waters. Tin [57, 58] has also been successfully separated.

6.1.3. Thin-Layer Adsorption Chromatography

In the separation of inorganic mixtures on thin layers, silica gel has been mainly used as the basic material for the thin layer. Because silica gel can function as an adsorbent as well as support for a stationary phase it is almost impossible to decide whether adsorption or partition is the operative mechanism.

Alumina may also be used for the preparation of thin layers for adsorption chromatography. Goller [585], using a solvent system prepared by mixing $0.5M$ potassium bromide in $1M$ hydrochloric acid with the same volume of 2-propanol, obtained good separations for mercury, cadmium, nickel, copper, iron(III), lead, silver and antimony(III). For the separation of halides, a mixture of 1-butanol, acetone, ammonia and water $(8:10:2:1)$ may be used. The separation is best carried out on a layer of a mixture of alumina and kieselguhr; R_F values increase in the order $IO_3^- < BrO_3^- < ClO_2^- < ClO_3^- < ClO_4^-$. Spraying with a mixture of aniline and phthalic acid will detect 5 µg of these ions [586]. For the separation of noble metals, circular chromatography on thin-layers has been proposed. It is simple and gives complete separation in less than two minutes. As solvent a mixture of acetone, acetylacetone and $2M$ hydrochloric acid $(100:10:3)$ was found suitable. Detection with rubeanic acid, thiourea and stannous chloride is suitable for down to 0.1 µg of all metals of the platinum group, gold and copper. Nickel, cobalt, manganese and iron interfere but they can be eliminated by preliminary separation [587]. A thin layer of alumina can also be used to separate ions in various oxidation states. Thus, for example, Galateanu [588] separated chromium(III) from chromate. Chromatography with a saturated solution of sodium sulphate gave sharp separation, chromium(III) remaining on the start, while chromate had an R_F of 0.5.

Lesigang and Hecht [589] proposed certain heteropolyacids as material for thin layers. They succeeded in separating caesium from lighter alkali metals, using water or silver nitrate solution for chromatographic development.

6.1.4. Organic Reagents as Adsorbents

The use of organic reagents as adsorbents in inorganic chromatography was first proposed by Erlenmeyer [59]. He used 8-hydroxyquinoline as an adsorbent for the separation of ions and he found that zones of metal oxinates formed according to the solubilities, the order being closely dependent on pH. The following sequence is found: VO_3^-, Mn^{2+}, WO_4^{2-}, Ag^+, Cu^{2+}, Bi^{3+}, Ni^{2+}, MoO_4^{2-}, Al^{3+}, Co^{2+}, Zn^{2+}, Mg^{2+}, Fe^{3+}, UO_2^{2+}, i.e. in order of solubility in water. The method is sufficiently sensitive for qualitative analysis. It permits, for example, detection of down to 2 µg of iron. Robinson [60] used it for the analysis of non-ferrous alloys, especially copper-nickel-zinc and copper-zinc-silver mixtures. Zinc was determined in these alloys, on a column of a mixture of equal parts of 8-hydroxyquinoline and starch. The solution for chromatography, the pH of which was adjusted to between 5 and 6, was introduced into the column and developed with 5% sodium acetate solution. Quantitative estimation was based on comparison of the height of the zone with a standard scale. For the determination of sodium and potassium he used violuric acid, the application of which was proposed by Erlenmeyer and co-workers [61, 62]. These authors used violuric acid either alone or mixed with silica gel, starch, etc., and by studying the adsorption of pairs of metallic ions found the following sequence of their sorption: K^+, NH_4^+, Mg^{2+}, Na^+; Ba^{2+}, Sr^{2+}, Ca^{2+}; Cu^{2+}, Pb^{2+}, Hg^{2+}.

Various organic reagents have been used, e.g. nitroso-R-salt for the determination of cobalt [26], dimethylglyoxime for nickel [63, 64], benzoinoxime [65], cupferron [66], dithizone, and others [25, 34, 67]. However, these methods have rather limited application. They may also be used for qualitative analyses of certain mixtures.

6.1.5. Adsorption Chromatography on Paper and Cellulose

The use of paper or cellulose as an adsorption material was investigated by Pickering [68−71, 742] and Lederer [72−80]. In contrast to partition chromatography on paper and cellulose, they studied the possibility of using water as eluent in separations of inorganic mixtures. Pickering studied the adsorption of hydrogen ions by measuring pH changes in solutions of $10^{-4}-10^{-2}M$ nitric acid. He determined the adsorption capacity of Whatman No. 1 paper and found it to be 10.8 μg of H^+ per g of dried paper. When measuring the adsorption of bivalent metals on paper he explained the separation process on the basis of ion-exchange, and found the exchange capacity for these ions to be 4 μmole per g of paper. The ion-exchange behaviour of paper in inorganic paper chromatography has also been investigated by Ackermann and Frey [646].

Lederer devoted his attention to the behaviour of inorganic ions during their separation on paper in aqueous solutions of mineral acids. He found that under suitable conditions, i.e. at certain concentrations, some heavy metal ions may be separated with hydrochloric or hydrobromic acid. Gallium, polonium, gold, rhenium, technetium, etc. could be separated on paper as chloro- or bromo-complexes. Mixtures of gold, gallium and iron can also be successfully separated by varying the concentration of hydrochloric acid used for development (gradient elution). Lederer also studied the possibility of sorption of some thiourea complexes of metals. He measured the R_F values of the complexes of platinum, gold, rhodium, ruthenium, osmium, bismuth, antimony, rhenium and technetium and compared them with the values obtained on chromatography with $2M$ hydrochloric acid in 1-butanol. He also investigated the separation of platinum metals by paper chromatography of their compounds with tin(II). Lederer also used adsorption paper chromatography for the study of complexes of chloroauric and bromoauric acid, and for the study of the chemistry of antimony pentachloride in aqueous solutions of hydrochloric, sulphuric and phosphoric acids. When studying the sorption of anions on cellulose he worked with

Table 6.2. Paper Adsorption Chromatography of Anions

Anion	R_F	Anion	R_F	Anion	R_F
WO_4^{2-}	0.95	TeO_4^{2-}	0.89	I^-	0.83
CrO_4^{2-}	0.64	BO_3^{3-}	0.87	SCN^-	0.77
MoO_4^{2-}	0.66	AsO_4^{3-}	0.98	$Fe(CN)_6^{4-}$	1.0
SeO_3^{2-}	0.87	AsO_3^{3-}	0.86	$Fe(CN)_6^{3-}$	1.0
TeO_3^{2-}	0.66	Cl^-	0.87	$Co(NO_2)_6^{3-}$	1.0
VO_3^-	0.54	Br^-	0.86	PO_4^{3-}	1.0

a sodium acetate and acetic acid solution as developer, and obtained the R_F values listed in Table 6.2.

The importance of these studies is not so much the development of separation procedures for these cations or anions, but the fact that they offer an overall picture of the behaviour of inorganic ions in paper and cellulose chromatography and that they stress those factors which are also important in paper partition chromatography.

6.2. Partition Chromatography

For inorganic analysis the methods of partition chromatography are the most important of all chromatographic methods. This is true both of paper chromatography, permitting qualitative and quantitative analysis of microquantities of separated substances, and of column chromatography on cellulose or silica gel. Paper chromatography was developed for the separation of various mixtures of inorganic compounds, and it is necessary that the vast number of results be arranged so that the necessary basic information can easily be found. To give full details of all the separations published would be unreasonable; instead R_F values of ions in a variety of solvent systems are given in the tables, accompanied by the references, enabling the reader to find further information in the literature. Chromatography on cellusose columns and silica gel is mentioned only briefly here, because these methods are most important in quantitative analysis and will be discussed later (see: Sections 7.2 and 7.3).

6.2.1. Paper Chromatography

Paper chromatographic methods are the most important in the chromatography of inorganic compounds, offering rapid and reliable identification of an unknown mixture of compounds, the separation of a certain ion from a known group of ions and, if necessary, its quantitative determination. In the case of inorganic mixtures it is usually impossible to determine unambiguously on the basis of the nature of the separated substances which solvent system or systems would be suitable for their separation. Therefore, in this section, the most important separations published in the literature will be reviewed, the solvent system given, and the general chromatographic behaviour of individual cations and anions briefly described. The R_F values of cations in published solvent systems are listed in the tables in the Appendix; R_F values of common anions are given in Tables 6.7 and 6.8. For the separation of noble metals and rare earths other solvents are more suitable than those usual for common cations, and the R_F values of these elements are given separately (Tables 6.3 and 6.4). Phosphorus and sulphur compounds (Tables 6.5 and 6.6) are also listed separately. In using the tables it is important to note the particular combination of elements separated with a given solvent system. The tables should serve as a rapid guide to the possibilities of separation of a particular element in a given combination with other elements. As a rule, the R_F values can be reproduced accurately only when all the conditions of separation given in the quoted literature are strictly observed.

Alkali Metals

The separation of alkali metals by classical analytical methods is rather difficult. In practice this group of elements must often be determined, but the conventional method of flame photometry is not always suitable. Therefore, great attention has been paid to the possibility of separating alkali metals by chromatographic methods. Various methods have been proposed, allowing more or less sharp separations. The majority recommend solvent systems containing alcohols as the main component. In such solvent mixtures, the R_F value increases with decreasing atomic number of the element. Usually it is not difficult to separate lithium or sodium and potassium from other elements of this group. One of the early successes of paper chromatography was Miller and Magee's separation of potassium, rubidium and caesium with concentrated hydrochloric acid — methanol — butanol — methyl isobutyl ketone (11 : 7 : 1 : 1) and of sodium from lithium with methanol-butanol (4 : 1) [434]. However, the separation of rubidium and caesium is usually not sharp, and with the majority of these mixtures separation is not feasible. A number of authors recommend lower alcohols with a larger or smaller water content for separations within the group. Low or negligible water contents are usually not well defined by authors, and the R_F values are very dependent on the particular mixture, and the values published by various authors for these mixtures are somewhat different. The separation can often be enhanced by combinations of alcohols and lower ketones. Monoethylcellosolve (ethylene glycol monoethyl ether) in a mixture with water and hydrochloric acid causes the order of alkali metals on the chromatogram to be reversed, i.e. lithium has the lowest R_F value, while caesium, rubidium and potassium have the highest [147]. However, the three last-mentioned elements cannot be separated with this solvent mixture either. Rubidium and caesium can be sharply separated with phenol saturated with $2M$ hydrochloric acid [221 − 223]. Good separation between lithium and sodium and potassium can also be obtained with alkaline solvent systems.

Pyridine, 2-propanol and acetic acid mixtures [143], or a methanol − ammonia mixture [225] permit the separation of these ions not only from each other but also from the majority of other ions, which remain on the start in the form of hydroxides. Good separation of all alkali metals can be made on papers impregnated with ammonium phosphomolybdate [215]. With the first solvent, 0.2M ammonium nitrate in 0.1M nitric acid, caesium and rubidium are separated from potassium, sodium and lithium. The chromatogram is then cut and caesium is separated from rubidium with 3.5M ammonium nitrate in 0.2M nitric acid, and sodium is separated from lithium with 95% ethanol. A satisfactory separation of all elements of this group may also be achieved with 0.1 or 1M hydrochloric acid on an asbestos paper [214, 218]. The separation of potassium, rubidium and caesium is also possible by centrifugal chromatography in 0.1 − 0.5M nitric acid mixed with 0.2 − 2M ammonium nitrate on papers impregnated with ammonium phosphomolybdate [520].

For quantitative determination of the alkali metals after their separation a comparison between the size of the spots with a chromatogram of standards is made, or the area of the spots is measured and then compared with values on a calibration curve [200, 224, 226]. Tristram and Phillips [195] determined the chlorides of the separated alkali metals by treating the chromatogram with silver nitrate, exposing it to direct light, reducing the AgCl by washing with a photographic developer, dissolving the metallic silver in nitric acid and titrating with thiosulphate. Vender [202] determined alkali metals by polarography after their chromatographic separation and detection with silver nitrate. Soós and Blazsek [203] proposed the determination of potassium and ammonium by indirect titration with sodium cobaltinitrate and for sodium and lithium indirect titration involving zinc uranyl acetate. Relatively accurate results were obtained by Hejtmánek and Hozmová [521] who determined rubidium and caesium as their picrates spectrophotometrically. Covello and Ciampa [522] determined alkali metals by flame photometry after chromatographic separation, potassium at 770 nm, caesium at 852 nm and lithium at 671 nm.

Alkaline Earth Metals

For the chromatographic separation of metals of the alkaline earth group, solvent mixtures containing lower alcohols are most suitable. Such mixtures are methanol and ethanol, methanol and hydrochloric acid, ethanol and acetic acid, methanol, 2-propanol and formic acid, or ethanol, propionic acid and ammonia, but a sharp separation can also be achieved by aqueous ethanol. The possibility of separation of alkaline earth metals among themselves, and also from a series of other elements, such as copper, zinc, cadmium, manganese, iron, cobalt and gallium, is offered by a mixture of tetrahydrofuran and hydrochloric acid. Precipitation chromatography based on the use of 8-hydroquinoline was used by Nagai [194] for a sharp circular chromatographic separation of copper, iron, nickel, cobalt, cadmium, manganese and magnesium. With methanol − ammonia, the complexing properties of EDTA can also be used for a good separation of calcium, barium and strontium [193]. For the separation of alkali and alkaline earth metals, Nakano and Shimada [192] proposed a two-dimensional arrangement. In the first direction, they separated both groups of ions in a mixture of ether, acetic acid and nitric acid (5 : 4 : 1). Then, for separation of the components of the two groups, they proposed methanol-nitric acid—hydro—chloric acid (8 : 1 : 1) or methanol—hydrochloric acid (9 : 1).

Imai [185] proposed inorganic solvents for the separation of calcium, strontium and barium. He first applied to the start one drop of $1N$ sodium sulphate solution, dried it, and then applied the mixture of ions to be separated. After development with an aqueous mixture of $1.3M$ ammonium thiocyanate and $1-2M$ sodium hydroxide he detected the zone at R_F $0.85-1.0$ with 8-hydroxyquinoline (Ca 0.95) and the zone of R_F $0.0-0.85$ with saturated aqueous sodium rhodizonate solution followed by 2% acetic acid (Sr 0.75; Ba 0.0). Good separation of calcium, strontium and barium is also obtained with $7M$ propanolic lithium nitrate solution [183]. Saxena and Singh [523] separated calcium, strontium and barium from one another and also from thorium with $8M$ aqueous ammonia, and calcium from beryllium and titanium [524] with $0.5M$ aqueous ammonia. The chromatographic behaviour of beryllium, magnesium, calcium, strontium and barium and also of aluminium and caesium, on Whatman No. 1 paper impregnated with dibutyl phosphate, with hydrochloric acid $(1-10M)$ as solvent, was investigated by Cvjetičanin [647]. It was shown that by choosing the appropriate concentration of hydrochloric acid and dibutyl phosphate, a good separation of alkaline earth metals from each other and from other cations is possible.

Paper chromatography is often used for the separation of radioactive substances and a number of papers deal with the separation of ^{90}Sr and ^{90}Y. Getoff [179] proposed a mixture of methanol, hydrochloric acid and citric acid; Kiba and co-workers [180] proposed a mixture of ethanol with 10% aqueous ammonium thiocyanate solution. Născuţin [188] recommended a mixture of methanol, acetic acid and water (40 : 1 : 60). Papers impregnated with Amberlite IR-120 (or IR-4B) and aqueous citrate solution as the mobile solvent are also suitable [186]. Stamm and Kohlschutter [182] used a paper impregnated with ferric hydroxide. Arnikar and Mehta [178] used an asbestos paper prewashed with hydrochloric acid, and $0.1M$ hydrochloric acid as the solvent. Strontium and yttrium can be separated successfully on a silica gel thin layer [187]. Silica gel G (Merck) and a solution of 0.2 mg of EDTA in 100 ml of methanol as solvent are convenient.

For quantitative determination of magnesium after its chromatographic separation several methods have been described. The simplest use comparison of spot size [150, 190, 199, 200]; planimetric measurement of the size of the spot gives more accurate results [197, 201]. When magnesium was determined in water the average error was less than 10%. For amounts of $0.01-10$ mg of alkaline earth chlorides Tristram and Phillips [195] used the same method as for alkali metals (see p. 83), based on the photographic process. Vender [202] determined magnesium polarographically together with calcium after washing the spots from the paper. After extraction of the separated metals from the paper, spectrophotometric methods [191, 206] or titrimetric methods [189] can be used. Direct photometry of paper chromatograms was proposed by Düsing and Kunze [184] who detected the barium and strontium spots with sodium rhodizonate and measured the absorption at 510 nm.

Arsenic, Antimony, Tin

Arsenic, antimony and tin usually display good mobility in solvents containing mainly butanol, with which they can be well separated from the majority of elements in the third analytical group and from each other. The R_F order is generally Sn > Sb > As, but if collidine with nitric acid is used for their separation, this sequence is reversed [124]. The addition of organic acids to the solvent system does not usually substantially change the mobility of the ions or the order of separation [125]. An interesting, though not too clear-cut, separation of antimony is described by Lederer [122] in one of his first papers on inorganic chromatography. Use is made of the strong sorption of weakly acidic antimony(III) solution on paper. Elution with water separates antimony from the majority of other ions present. By paper chromatography antimony(III) can be well separated from antimony(V). The best solvent is a mixture of methanol, ethanol and hydrochloric acid. A number of solvent systems also permit the separation of arsenic(III) from arsenic(V). Thus, e.g. a mixture of acetone, methyl isobutyl ketone (MIBK) and aqueous nitric acid gives not only a sharp separation of both oxidation states of arsenic, but also their separation from iron, copper, nickel, cobalt, bismuth, uranium, thorium, and mercury [105]. A mixture of methanol and ethanol with aqueous hydrochloric acid is suitable for the separation of the oxidation states of arsenic, and also for the separation of oxidation states of copper, antimony, chromium, and molybdenum [110]. The best separation for bivalent and quadrivalent tin was obtained with ethyl acetate − methyl salicylate (1 : 1) as solvent [648]. The R_F values are 0.31 and 0.97 for Sn(II) and Sn(IV) respectively. To demonstrate the presence of arsenic by paper chromatography, use can be made of the fact that on chromatography in a mixture of acetone, acetylacetone and hydrochloric acid, arsenic gives two spots, one at the solvent front, and the second at R_F 0.55. Neither of these spots coincides with those of antimony and tin [113]. In organic material, arsenic was detected by means of the fact that in a mixture of methanol and sulphuric acid it remains on the start, while numerous other ions, e.g. mecury, copper, zinc, and thallium, move at various rates with the mobile phase. When chromatographed in butanol, arsenic remains on the start, a fact which has been used for its determination in glass [114]. Murata [115] investigated the separation of a mixture of arsenic, selenium and tellurium, and developed a method for the separation of compounds of these elements and also of their different oxidation state. For the separation of arsenic in the analysis of various inorganic mixtures, a mixture of 1-butanol, hydrochloric acid and water can be used with advantage. It separates arsenic from copper, lead, cobalt, nickel, iron, and manganese [102]. Complex sulphides of arsenic, antimony, and tin, dissolved in ammonia, can be separated sharply with $0.6M$ aqueous ammonia; antimony remains on the start, arsenic has R_F 0.75 and tin 0.86 [103]. Tin(II) and antimony(III) can be separated from the majority of other metal ions − in a formic acid − cyclohexanol mixture (3 : 7). Tewari [112] was able to separate arsenic, antimony and tin with 1-butanol saturated with $3M$ hydrochloric acid and he used this method for the detection of these ions in toxicological analysis. Chromatographic methods are also suitable for the control of inorganic syntheses. A good example of such an application are the papers by Thil and co-workers who applied partition paper chromatography for the control analysis of condensed phosphates and arsenates [116, 117].

Janjić and Ćelap used papers impregnated with $0.05M$ lead nitrate for the separation and semiquantitative determination of arsenates, phosphates and chromates by precipitation chromatography [118]. Labić [119] separated arsenic, antimony and tin on papers impregnated with thiosulphate and iodide, or with thiosulphate and oxalate.

Direct photometry can be used for determination of arsenic after the separation on paper. Vaeck [113] recommends a determination based on the use of rubeanic acid. Heyndryckx [114] recommends spraying the chromatogram a buffered solution of 8-hydroxyquinoline and direct densitometric determination. For determination of antimony the spots are best eluted from the paper and the eluate is submitted to direct photometry. For determination of tin, colorimetric or polarographic methods have been proposed [99, 100]. An accurate determination of tin in antimony, with a silica gel column, has been described by Šulcek and co-workers [58].

Beryllium

Beryllium is often separated from alkaline earth metals or from aluminium chromatographically. It is best separated from alkaline earth metals with alcoholic solvents. Beryllium can be sharply separated from calcium, strontium, and cerium with a mixture of methanol and hydrochloric acid; the separation from magnesium is not so good [141]. The same is true if a mixture of ethanol and acetic acid is used [145]. If pyridine or collidine is added to the alcohol mixture the separation is improved and a sharp separation of beryllium from magnesium becomes possible, e.g. with a mixture of 2-propanol, pyridine and acetic acid, beryllium moves closely behind the solvent front, but the mobility of magnesium is approximately half that of beryllium, and the mobility of other alkaline earth metals is lower still [143]. This mixture also gives sharp separation from aluminium, but does not separate beryllium from copper, cobalt, nickel, iron or cadmium. The separation from alkali metals is very good. The latter remain near the start, except for lithium (R_F 0.61). A collidine—methanol—1-butanol—acetic acid mixture gives similar results [150], but separation from aluminium is impossible with this solvent. Aluminium is separated from beryllium by a mixture of ethyl methyl ketone and hydrochloric acid [151]. Aluminium remains near the start, while beryllium has R_F 0.51, and iron moves with the solvent front. Satisfactory results are obtained when beryllium is separated from aluminium with a mixture of 1-butanol and hydrochloric acid [144, 146, 152]. Carvalho and Lederer [149] used this solvent system for the separation of beryllium, indium, germanium and gallium. By varying the concentration of the acid in the solvent system they determined suitable conditions for partial separations. A good method for the separation of beryllium from the majority of other ions is a combination of precipitation with ammonia from a medium containing EDTA, and subsequent chromatographic separation of beryllium, uranium and titanium, with a mixture of 2-propanol, acetylacetone and hydrochloric acid [148]. Beryllium can be separated from aluminium, iron, copper, cobalt, manganese and zinc by paper chromatography of the EDTA complexes [489].

After chromatographic separation beryllium can be determined semi-quantatively or quantitatively by planimetry or by photometry. Photometry is more accurate than measuring the size of the spots which has a 15% relative error [148]. Direct photometric determination of beryllium on paper chromatograms has been described by Filatova who separated beryllium and aluminium on paper impregnated with 8-hydroxyquinoline. With a fluorimetric finish the sensitivity of her method was 0.9 µg of Be in one drop of the analysed solution. For indirect photometric determination, beryllium may be eluted from the paper with 1% sodium hydroxide solution and determined with quinalizarin [146]. Ader and Alon [153] obtained good results when determining beryllium in ores by using Eriochromocyanine R for the detection and comparing the spots with those on a standard chromatogram. A similar method was used by Agrinier, who used quinalizarin for the detection [154].

Gallium

A problem often encountered in the determination of gallium is its separation from larger quantities of aluminium. The problem can be solved by paper chromatography with a mixture of phenol, methanol and concentrated hydrochloric acid [241], in which gallium moves more than six times faster than aluminium. The separation is sharper still with 1-butanol saturated with $1-12M$ hydrochloric acid [280]; aluminium remains near the start but the mobility of gallium increases with increasing concentration of hydrochloric acid. An interesting separation is described by Beckmann and Lederer [74]. Using descending chromatography on Whatman No. 1 paper they separated some heavy metals with hydrochloric or hydrobromic acid or chloride solutions of suitable concentration. This separation is based on the adsorption of metal ions and their separation in the form of chloro- and bromo-complexes. Good results were obtained for gallium with $7M$ hydrochloric acid, lithium chloride, $7.5M$ calcium chloride, etc. Olenovič and co-workers [527] used the EDTA complexes to separate gallium, indium, cadmium, and zinc. With a mixture of 1-butanol—water—concentrated hydrochloric acid—EDTA (450 ml + 120 ml + 9.2 ml + 0.2 g) they separated these cations in the following order of increasing R_F: Ga < In < Cd < Zn. Gallium can be separated from indium, zirconium and uranium in the form of the mandelic acid complexes, with 1-butanol as solvent. Gallium moves with the solvent front and the order of R_F values of the other ions is In > > Zr > UO_2^{2+} [526].

Gallium can be determined by elution from the paper and photometry of the eluate after treatment with Rhodamine B or Malachite Green, or fluorimetry with 5,7-dibromo-8-hydroxyquinoline [275].

Germanium

For the separation of germanium from other elements mixtures of butanol and mineral acids are most suitable. Lederer [149, 273] proposed 1-butanol saturated with $1M$ hydrochloric acid. Ladenbauer [274, 275, 277, 278] and co-workers preferred 1-butanol saturated with hydrobromic acid. Good separation was achieved by Nagy and Pólyik using a mixture of chloroform, ethanol and hydrochloric acid [276]. Germanium may be determined semi-quantitatively on chromatograms from the spot sizes. To within $\pm 6\%$ the area of the spot is a linear function of the logarithm of concentration [278]. However, photometric determinations are more common and more precise. Phenylfluorone is usually used as reagent [277, 279]. The spots can be evaluated either directly on the chromatogram or after elution with water. It is important to make sure that bismuth, if present, is well separated during the chromatographic process [277]. Agrinier [99] used alcoholic solutions of Gallion and haematoxylin for detection, and by this method determined germanium in minerals.

Aluminium

For the chromatographic separation of aluminium on paper, mixtures of butanol and hydrochloric acid are most commonly used. In such mixtures, especially with increased concentration of acid, numerous cations, e.g. iron, cobalt, copper, zinc, cadmium, mercury, beryllium, bismuth, can be separated from aluminium, which remains near the start. This behaviour has been used by numerous workers for the separation of aluminium from iron in mineral waters, or of beryllium and aluminium in beryl [146], and for the determination of aluminium in beryllium or *vice versa* [152], as well as in other materials. For the separation of copper a mixture of ethyl methyl ketone—hydrochloric acid-water was used [283]. Gallium, indium and thallium can be well separated from aluminium with a mixture of phenol, methanol and hydrochloric acid [241, 287]. Increasing the proportion of methanol and hydrochloric acid in this mixture can substantially increase the R_F value of aluminium but the separating power diminishes appreciably. Aluminium has medium mobility in a mixture of tetrahydrofuran and hydrochloric acid [205] which provides good separation. In detailed studies on the application of paper chromatography in qualitative and quantitative microanalysis, Lacourt [288—292] and co-workers devoted great attention to the separation of aluminium. They investigated the separation of iron, titanium, aluminium, vanadium, nickel and cobalt, and demonstrated the possibility of direct and indirect quantitative methods on the basis of these separations. Qureshi [489] has examined the EDTA complex.

For determination of aluminium, photometric methods are most commonly used. For direct determination Aluminon is most suitable [282] and for indirect photometry after the elution of aluminium with hot water acidified with hydrochloric acid, Chromazurol has been proposed [285]. Magee and Scott [287] developed a procedure for uranium, aluminium, gallium, indium, and thallium in which the oxinates were separated with a mixture of phenol and methanol. The spots of the coloured oxinates were compared with standards and the amount of the metals determined semiquantitatively. For a more precise evaluation, the oxinates may be eluted from the paper with organic solvents and determined photometrically. Aluminium and titanium, etc. in ferric oxide can be separed by circular chromatography using a mixture of 1-butanol, ethyl acetate and hydrochloric acid and determined on the basis of the linear relation between the zone width and the logarithm of the concentration.

Chromium

Chromium(III) is not very mobile in solvents containing butanol. This can be used for the separation of chromium(III) from ions which move faster in these solvents, such as Bi^{3+}, Cd^{2+}, As^{3+}, As^{5+}, Sn^{2+}, Sn^{4+} and Zn^{2+} [156]. The addition of acetylacetone allows separation from iron but does not increase their mobility [212]. The mobility of chromium increases substantially in a mixture of butanol, acetic acid and acetoacetic ester, thus permitting further separations, e.g. from aluminium, cobalt, nickel, manganese, etc. A mixture of collidine and nitric acid, in which chromium remains on the start, is suitable for separation of chromium from copper [124, 212]. In mixtures of lower alcohols with mineral acids, mostly hydrochloric acid, the mobility of chromium is characterized by medium R_F values. By use of suitable alcohols and hydrochloric acid concentrations various other separations can be achieved [155, 241]. Dioxan — nitric acid — water (200 : 2 : 5) has been used to separate chromium from copper, zinc, cadmium and nickel [698].

Chromatographic methods can be used also for the separation of the various oxidation states of chromium, e.g. with a solvent containing diethyl ether and concentrated hydrochloric acid, chromium(III) can be sharply separated from chromium(VI) [296], and with a mixture of 1-butanol, acetic acid and ethyl acetate, chromium(VI) is separated from chromium(III) and chromium(II) [297]*.

Another method of chromatographic determination of chromium is based on the reaction of chromate ions with silver and on the limited solubility of silver chromate in acetic acid [299, 528]. This method gives a semiquantitative determination, as the area of the silver chromate spots is proportional to the amount of chromate present. Indirect photometric determination after the elution of the spots is however more precise. At relatively high concentrations of chromium direct photometry using the chromate or dichromate colour may be carried out; at lower concentrations the reaction with diphenylcarbazide is used. Small amounts of chromium in the presence of manganese, cobalt, iron, and nickel can be determined according to Fojtík and Koprda [529, 530], by means of radioactive ^{51}Cr and planimetry of radiographs. Radioactive chromium was used for labelling by Galateanu [531] in a study on the chromatographic separation of chromium(III) from chromium(VI).

---

* This ignores the fact that Cr(VI) and Cr (II) cannot coexist! Isomers of chromium (III) compounds can be separated chromatographically with methanol–acetic acid (9 : 1) [298].

Indium

Often the separation required is of indium from aluminium, germanium, gallium and thallium. In most cases 1-butanol saturated with hydrochloric acid of various concentrations has sufficient separating power [149]. For the separation of indium from cadmium Olenovič [293] recommends the addition to the solvent of 0.2% EDTA solution. Fişel and co-workers separated the elements mentioned above with sufficient sharpness by using pentanol saturated with $4M$ hydrochloric acid [282], or mixtures of phenol and methanol, or propanol and concentrated hydrochloric acid. A mixture of phenol and methanol is also recommended by Magee and Scott [287]. Gallium, zirconium, and uranium can be separated in the form of their complexes with mandelic acid [526].

For quantitative determination indirect spectrophotometric methods may be used. Indium can be determined with sufficient accuracy in eluates from paper chromatograms, by spectrophotometry at 515 nm, after its reaction with diphenylcarbazone [282]. For direct determination on paper, Magee and Scott [287] proposed detection with oxine and semiquantitative determination of the separated amount by comparison with standard chromatograms. The metal oxinates can also be eluted from the paper and determined spectrophotometrically.

Cadmium

Chromatographic methods are widely used in the determination of cadmium compounds. In the majority of solvent systems, cadmium is relatively very mobile, giving good separation from most mixtures. With butanol-based solvents containing low concentrations of hydrochloric acid, cadmium can be separated from iron, aluminium, chromium, manganese, nickel, and cobalt, and also from copper and lead. It may be separated from bismuth with a mixture of dioxan, antipyrine and nitric acid [101]. Separation of cadmium from zinc is difficult, as zinc has a very similar chromatographic behaviour. However, their pyridine—thiocyanate complexes can be separated with a mixture of ethanol, acetone, sodium bromide and ammonium acetate [157], but separation of the ammine complexes is simpler. With a mixture of methanol and ammonia, cadmium is separated from zinc, and also from cobalt and copper [164]. Both elements can also be sharply separated with a mixture of 1-butanol, ammonia and pyridine [161]. Erdem [363] showed, however, that in the separation of cadmium by a solvent containing ammonia, well-separated spots were sometimes formed, and he attributed this to the formation of various ammine complexes. Solvents containing mainly chloroform as well as alcohols or ketones and hydrochloric acid have been employed for the separation of copper, cadmium, and bismuth by circular chromatography [532].

For the separation of cadmium from some other elements, complexes with certain organic reagents can be used. Nagai [255, 256], Minczewski [170], and Venturello [165] studied the separation of mercury, copper, bismuth, cadmium, lead, zinc, and silver dithizonates under various conditions and in many different solvent systems. The order of the separation depends primarily on the solubility in the solvent used. Jursík [359] separated the glycine complexes of cadmium, copper, nickel and zinc, with a mixture of ethanol or 2-propanol and water. Nagai recommends precipitation chromatography with thioacetamide [257] for the separation of mercury, copper, bismuth and cadmium, and for another, similar, group of elements precipitation chromatography with 8-hydroxyquinoline [194, 253]. Singh and Dey proposed separating copper, nickel, cobalt, iron, and cadmium in the presence of oxalate [246], citrate [245], or tartrate [244]. Ziegler and Winkler [361] demonstrated that in solvents containing hydrogen sulphide (1-butanol saturated with $3M$ hydrochloric acid and gaseous hydrogen sulphide) copper, lead, bismuth and mercury remain on the start, while cadmium separates as a distinct zone. For certain separations, paper impregnated with 4-hydroxybenzothiazole [360] has been applied successfully. Cadmium has been separated from copper and mercury with 1-butanol saturated with $3M$ hydrochloric acid, and from copper, zinc, chromium and nickel with dioxan — nitric acid-water (200 : 2 : 5), when these metals are present in industrial effluents from e.g. the plating industry [698].

For semiquantitative determination of cadmium and bismuth after their separation with 1-butanol — water — acid solvent Popper and co-workers suggested planimetric evaluation of the spots [533].

In certain circumstances, the reaction of cadmium with dithizone may be used for a quantitative photometric determination. The best method, however, as the sensitivity is very high, is oscillographic polarography, after elution from the paper [364].

Cobalt

Many different methods have been developed for the separation of cobalt by paper chromatography. As well as the separation of different oxidation states, the mobilities of numerous complexes formed by cobalt have been studied and many methods proposed for its separation from certain mixtures and its subsequent quantitative determination. Very often the formation of the chloro-complex is used for separation. In butanol or acetone containing a small amount of hydrochloric acid, the mobility of cobalt is very low, but increases with increasing concentration of the acid, and it can be separated completely from nickel, which even at higher concentrations of acid remains near the start. The separation of cobalt from nickel, and also from iron and copper, is the most widely studied chromatographic problem with this group of elements. They can also be separated with mixtures of tetrahydrofuran and ether or hydrochloric acid [205], benzene and acetic acid [235], methanol, dioxan and cyclohexane [236], acetic acid, pyridine and hydrochloric acid [335], and also solvents containing methylamine [534]. One separation of cobalt from nickel, where the former moves with the solvent front and the latter remains on the start, is done with a mixture of acetone, ethanol and hydrochloric acid. In all these solvents cobalt moves faster than nickel. The opposite order is observed in solvents containing ammonia, e.g. in methanol-water-ammonia, cobalt remains near the start, while nickel has R_F about 0.60. Hahn and co-workers [239] observed an interesting effect during the separation of cobalt from copper in a mixture of butanol, acetic acid, ethyl acetate and water. Cobalt formed three distinctly separated spots which, evidently, should be attributed to complexes solvated to various degrees.

Considerable attention has also been given to the separation of complex ions of cobalt. Singh and Dey investigated the possibility of separating cobalt, nickel, iron and copper as oxalates [246, 344], citrates [245, 345, 246], and tartrates [340]. Under suitable conditions they obtained good results. For the separation of copper, nickel, cadmium, zinc, and cobalt they used a mixture of ethanol with various concentrations of ammonium chloride and ammonia. It was found advantageous first to separate cobalt together with uranium, palladium, thorium and zirconium by precipitation with 1-nitroso-2-hydroxy-3-naphthoic acid and then to separate the complexes with a mixture of acetone, isobutyl alcohol and hydrochloric acid [272]. Lederer [348] studied the chromatographic behaviour of certain cobalt(III) complexes, e.g. hexa-amminocobalt(III)chloride, and *trans*-dinitrotetra-amminocobalt(III) chloride. Stefanović and Janjić investigated the separation of *cis* and *trans* isomers of dichloro-bis(ethylenediamine)cobalt(III) chloride and the similar iodide complex [349]. Nagai separated a mixture of cobalt, nickel and copper [350—353] on papers impregnated with 8-hydroxyquinoline [194, 249, 251, 253] and rubeanic acid [252]. Ashizava [354] and Nagai [535] determined traces of cobalt on papers impregnated with dithizone. Imai [169] separated manganese, cobalt, nickel, and zinc by impregnating the paper with 1-nitroso-2-naphtol in the $0.1-0.2$ R_F region and with dimethylglyoxime in the $0.4-0.5$ R_F region. Manganese remains on the start when run in an ammonium chloride solution, cobalt and nickel are retained in the first and second impregnated zones respectively while zinc moves with the solvent front. Impregnation of the chromatogram with 1-nitroso-2-naphthol can be used for the quantitative determination of cobalt when eluting with water, by measuring the spot size [355]. Chromatography of the EDTA complex of cobalt has been studied by Qureshi [489].

Cobalt can be determined photometrically directly on the paper by measuring the spots with a densitometer, or by elution and subsequent determination with a suitable reagent. For direct methods 1-nitroso-2-naphthol is the most suitable reagent [343, 356]. After elution other reagents can also be used, e.g. nitroso-R-salt or α-fural monoxime. Planimetric estimation of the spot sizes is only suitable for semiquantitative determinations. According to Ota [536] the spot area is proportional to the logarithm of the concentration of cobalt.

Manganese

There are numerous methods of separating manganese by paper chromatography. In general, manganese has a very low mobility in the majority of solvent systems. In 1-butanol-hydrochloric acid it remains near the start, but if a lower alcohol is used, higher R_F values are observed. Manganese can be separated from iron, bismuth, copper, cadmium and zinc with such solvents, but the separation from cobalt and nickel is usually impossible. However, manganese can be separated from nickel and cobalt with several other solvents, such as pyridine, acetic acid and hydrochloric acid [213], ethyl methyl ketone and hydrochloric acid [120] and methanol, 2-propanol, acetic acid and hydrochloric acid [329]. Distinct separation of manganese from cobalt and nickel can be obtained in a mixture of ammonium chloride and ammonia. Manganese remains on the start, while cobalt and nickel move with the solvent front [328]. Solvents containing pyridine or collidine [212] are also suitable for the separation of manganese. They generally give a sharp separation from bismuth, tin, aluminium, chromium and iron. For the separation of manganese from iron, nickel, cobalt and copper, Berg proposed separating the 2-thenoylperfluorobutyrylmethane or 2-furoylperfluorobutyrylmethane complexes, using a mixture of light petroleum and methanol, or dioxan [248]. Sharp separation from the majority of accompanying elements is also possible with tetrahydrofuran and hydrochloric acid [205]. Qureshi has investigated the chromatography of the EDTA complex [489]. Nagai [194, 253] separated copper, iron, nickel, cobalt, cadmium, manganese and magnesium oxinates; Yokota and Shimada [258] used paper impregnated with kojic acid and separated the metals with a mixture of 1-butanol and acetic acid, in the following order of decreasing R_F: Fe $\rightleftharpoons$ Cu $\rightleftharpoons$ Mn $=$ Ni. Imai [169] proposed inorganic solvent mixtures for the separation of cobalt, nickel, manganese and zinc; e.g. ammonium chloride, thiocyanate, and ammonium nitrate or sulphate (*cf.* cobalt, p. 92). Shukla [168] recommends centrifugal chromatography for the separation of metals of the third group.

For quantitative determination of manganese, indirect photometric methods are best. The colour of the permanganate ion may be used, or radiometric methods can be applied, using radioactive ^{54}Mn and planimetric evaluation of radiographs [529, 530].

Copper

Analytically copper is one of the most studied elements, for the simple reason that it is one of the elements that form very many complexes. Many methods for its separation and detection have been proposed, and more organic reagents have been proposed for its photometric determination than for any other element. Chromatography offers a number of possibilities for the separation of copper from various mixtures under various conditions. In the commonest solvent system for inorganic chromatography (1-butanol containing hydrochloric acid) the mobility of copper is dependent on the acid concentration. It varies between R_F 0.1 in 1-butanol saturated with $2M$ hydrochloric acid, and R_F 0.65 in 1-butanol–concentrated hydrochloric acid (7 : 3). As the mobility of other elements also varies with solvent composition, changes in the proportions of the components of the solvent mixture can be exploited for sharp separation. Formation of ammine complexes and separation with 1-butanol saturated with $2M$ ammonia is very useful. With this solvent a number of elements remain of the start (as hydroxides), while copper moves considerably (R_F 0.54). If a mixture of 50% aqueous ethanol and ammonia or ammonium chloride is employed, copper remains on the start at low ammonia concentrations (R_F 0.06), but moves quite appreciably when the concentration of ammonia is increased and ammine complexes are formed. The formation of ammines can be used for the separation of copper, nickel, cadmium, zinc, cobalt and silver [34]

Several solvent systems based on lower aliphatic alcohols may be used, e.g. copper can best be separated from zinc, cadmium, bismuth, silver, gold, iron, and many other elements by a mixture of ethanol and hydrochloric acid. Similar mixtures, such as 2-propanol and hydrochloric acid, or a mixture of ethanol, 2-propanol and hydrochloric acid, can also be used. Good separation is possible in solvents containing methanol. If these contain hydrochloric or sulphuric acid, copper can be successfully separated from the alkaline earth metals, and if the solvent also contains dioxan and cyclohexane, cobalt and nickel can be separated. Cobalt, nickel, and copper can also be successfully separated chromatographically with tetrahydrofuran–concentrated hydrochloric acid, the order of R_F values being Cu > > Co > Ni, and also with acetone–hydrochloric acid mixtures, in which the sequence depends on the concentration of hydrochloric acid. In acetone with $6M$ hydrochloric acid (95 : 5) the same R_F sequence is found, Cu > > Co > Ni, but if the concentration of the acid drops to $0.5M$, the sequence is changed to Co > Cu = Ni. In some solvent systems, copper(I) and copper(II) can be separated (e.g. in acetone and glacial acetic acid). Mixtures of alcohols with traces of hydrochloric acid and with ketones or ethyl acetate generally give optimum separation of copper(I) and copper(II) [649]. With solvents containing diethyl ether and nitric acid, copper can usually be well separated from lead. Thielemann has separated copper from cadmium, mercury, zinc, chromium and nickel in analysis of industrial effluent from plating plant [698].

The formation of complexes with organic ligands is often employed in the separation of copper. Singh and Dey [244−246] in their study on the separation of copper, cadmium, iron, nickel, and cobalt, made use of the complexing ability of tartrates, oxalates and citrates, using methanol as the solvent. Berg and Strassner [238] chromatographed acetylacetonates of cobalt, copper, and nickel and found that separation is best in a mixture of cyclohexane, dioxan and methanol (84 : 10 : 6). Qureshi examined the chromatography of the EDTA complex [489]. Fişel and Giurgiu [247] studied the chromatographic behaviour of phenoxyacetates of manganese, cobalt, nickel, copper, zinc, and cadmium in the presence of phenoxyacetic acid or in the presence of certain bases such as ammonia, pyridine, or ethylenediamine, and succeeded in separating pairs of these metals. Kaursz and Orbán [157] investigated the pyridine−thiocyanate complexes of cadmium, zinc, cobalt, nickel, and copper, and described their separation with a mixture of ethanol, acetone, sodium bromide and ammonium acetate. In a study of the chelates of iron, copper, nickel, cobalt, and manganese with 2-thenoyltrifluorylacetone, Berg and McIntyre [235] chromatographed them on paper, cellulose, starch, silica gel and alumina, and in all cases obtained good separation. They also successfully used certain β-diketone chelates, especially 2-thenoylperfluorobutyrylmethane and 2-furoylperfluorobutyrylmethane complexes [248]. Precipitates with organic reagents were employed for separations by Nagai. He separated cobalt, copper, and nickel on papers impregnated with rubeanic acid [252], and on papers impregnated with oxine he separated copper, iron, nickel and cobalt, using ammonia or a mixture of ammonia and organic reagents [249, 251, 253, 254]. For the separation of iron, copper, zinc, mercury and uranium he used papers impregnated with cupferron [167]. Copper, mercury and bismuth were separated as sulphides on papers impregnated with thioacetamide [257]. In a second series of papers Nagai used organic reagents which give soluble compounds with the separated ions. Mercury, copper, bismuth, cadmium and lead were separated in the form of their dithizonates, with a mixture of $0.1M$ nitric acid and acetone (10 : 1) [255, 256], and nickel, cobalt and copper with alcoholic ammonia on papers impregnated with dithio-oxamide [250]. Yokota and Shimada separated copper and iron from manganese and nickel, using paper impregnated with kojic acid [258]. The successful separation of *cis* and *trans* isomers of copper glycinates in solvents containing lower aliphatic alcohols or ketones and water [537] represents a good example of the use of paper chromatography to solve inorganic chemistry problems.

For chromatographic determination of copper various procedures have been developed. Among the indirect methods those using titrimetry, photometry or polarography are best. Electrolytic determination has also been proposed [237]. Of the direct methods, planimetry is simplest [171, 227], although according to Boichinova and Aleskovskii, [228] not always suitable, and photometry is the most convenient. For direct photometric determination of copper spots on paper chromatograms, rubeanic acid [113, 259], Zincon, α-benzoinoxime [230], etc. have been employed. A detailed study of quantitative evaluation of paper chromatograms was published by Lacourt [173, 231, 232]. When separating cobalt, copper and zinc, she checked the accuracy of quantitative determination by direct photometry, and was able to determine the separated ions with $3-4\%$ error. When investigating titrimetric methods she found that often it is not necessary to elute the substance from the paper before the titration. For the determination of copper it is sufficient to cut out the spot and immerse it in an ammoniacal solution, add Murexide, and titrate with EDTA until a red colour appears. Smoczkiewiczowa and Mizgalski [229] used spectrographic analysis of the ash after incineration of the spot in a silver crucible which also served as electrode. Yamada [230] studied the possibility of using retention analysis for the determination of copper on paper chromatograms and obtained satisfactory results with a solution of α-benzoinoxime in a mixture of butanol and water.

Molybdenum

Molybdenum can be separated fairly well by paper chromatography using 1-butanol saturated with $1M$ hydrochloric acid, from common elements such as iron, copper, cobalt, chromium and nickel, and also from titanium and uranium. More often, however, molybdenum is required to be separated from vanadium, tungsten, titanium, zirconium, niobium, tantalum, uranium, etc. and for this purpose, solvent systems based on alcohols and hydrochloric acid are suitable. The separating power of the solvent mixture can be varied within certain limits by changing the concentration of the acid in ethanol and sufficiently sharp separation of molybdenum may be obtained [160, 266]. However, a mixture of chloroform, ethanol and hydrochloric acid is more convenient, because in it molybdenum has high R_F values (0.82), while vanadium, tungsten, zirconium, uranium, thorium, and titanium, remain either on the start or close to it [276]. A mixture of methanol, ethanol and hydrochloric acid separates the oxidation states of a number of elements, including molybdenum [110]. The compounds of molybdenum(V) have R_F 0.2, while those of molybdenum(VI) have R_F 0.55. Molybdenum can be separated from vanadium and chromium with a mixture of acetone and hydrochloric acid (possibly with chloroform present as well) [364], and also with a mixture of pentanol and formic acid [306].

For photometric determination of molybdenum after its elution from the paper, the reactions with thiocyanate, dithiol, or thioglycollic acid are most suitable. A quantitative determination of molybdenum by ascending paper chromatography is described by Kielczewski and Uchman [538]. The areas of the white spots of lead molybdate formed on papers impregnated with 0.002N lead sulphate are proportional to the quantity of the molybdenum present.

Nickel

Nickel usually has to be separated from cobalt, copper, iron and manganese. These separations have already been described in the section on cobalt (see p. 91). It has been separated from other metals in effluents from plating plant [698].

For quantitative determination of nickel direct methods [228] can be used, e.g. planimetry (which is not always suitable), or direct photometry at 632 nm after detection with rubeanic acid [343]. The latter method is more suitable. After elution from the paper, nickel may also be determined in microgram amounts by titration with EDTA [367], oscillopolarographically [358], polarographically, or by a suitable spectrophotometric method. Trace determination of nickel by combining paper chromatography with radiometric precipitation was described by Chiotis and co-workers [650]. After paper chromatographic separation with acetone — concentrated hydrochloric acid — water (87 : 8 : 5), nickel is located as the blue complex with Alizarin Red S and ammonia. The nickel zone of the chromatogram is then immersed for three hours in a solution of ammonium sulphide (labelled with ^{35}S) to precipitate Ni^{35}S, which is located by autoradiography, removed and counted by scintillation.

Niobium and Tantalum

For the separation of niobium and tantalum solvent systems containing certain ketones and hydrofluoric acid are suitable. The use of mixtures of hydrofluoric acid with ethyl methyl ketone (85 : 15) in which niobium and tantalum move with the solvent front, permits separation from a number of other ions [130]. Methyl isobutyl ketone or diethyl ketone mixed with hydrofluoric or nitric acid, separate niobium from tantalum. In both systems the R_F value of tantalum is the higher. A mixture of ethyl methyl ketone with hydrochloric acid also satisfactorily separates niobium from tantalum, but the order is reversed. A mixture of ethyl methyl ketone and hydrofluoric acid was used by Williams and co-workers [132—136] for this separation on a cellulose column. They developed procedures for the determination of niobium and tantalum in minerals. The semiquantitative determination of these elements was studied by Agrinier [137]. A mixture of ethyl methyl ketone and hydrochloric acid of various concentrations is suitable for the separation of niobium, tantalum and titanium [138]. Blasius and Czekay [128] recommend using the difference in mobility between the oxalate complexes of niobium and tantalum to obtain a sharp separation with solvents containing acetone or 1-butanol. The separation of oxalate or tartrate complexes in acetylacetone or dioxan mixed with hydrofluoric acid has been used for the separation of small amounts of hafnium and niobium from large amounts of tantalum [539]. 1-Butanol mixed with an aqueous solution of nitrates can be used for the separation of ^{95}Nb from ^{95}Zr. The solvent composition $7M$ LiNO$_3$ $-2M$ HNO$_3$ $-$1-butanol $(1:1:4)$ is the most suitable.

Niobium and tantalum can be determined quantitatively after chromatographic separation, either by comparing the size of the spots with standards [130, 140], or by colorimetric methods. After detection with 8-hydroxyquinoline the niobium and tantalum spots can be eluted from the paper by shaking the paper sections with $2M$ hydrochloric acid, and the oxinates then extracted with chloroform. The optimum pH for niobium is 8.5, and for tantalum 10. Photometric measurements are made at 400 nm. Martin and Magee [131] recommend cutting out the spots after detection, fusing with potassium hydrogen sulphate, extracting the melt with an oxalate solution, and determining photometrically after the addition of pyrogallol.

Lead

The behaviour of lead is characterized by its low mobility in the majority of solvents. With many, lead remains on the start. This is especially true of solvent systems containing sulphuric acid which precipitates lead as the sulphate, and of alkaline systems based on pyridine [83] or collidine [81]. Good separation can be achieved with dilute sulphuric acid alone [85]. If the concentration of the acid is low [0.05N], lead remains on the start, but with increasing acid concentration its R_F value also increases (0.17 with 0.2N acid). Accompanying elements, such as silver, mercury, copper and cadmium move either faster or even slower. Zero R_F values for lead are also observed for solvent mixtures with acetone, acetic acid, tartaric acid or methyl acetate [92], and also in 1-butyl alcohol saturated with $1M$ hydrochloric acid [104]. However, with increasing concentration of hydrochloric acid in the last-mentioned mixture the R_F value of lead increases, to attain R_F 0.52 at an acid concentration of 30%. Solvents precipitating lead at the start usually allow its separation from the majority of common accompanying elements such as copper, zinc, cadmium, bismuth, etc. However, in view of further manipulation of the chromatogram it is sometimes undesirable to transform lead into an insoluble form. In such a case it can be separated from common accompanying elements by using mixtures of acetone and butanol, hydrochloric acid and acetylacetone [162], 2-propanol, acetic acid and water [86], and 1-butanol, acetone, nitric acid and acetylacetone [98].

A consecutive separation of mercury, copper, cadmium, bismuth and lead has been developed by Imai [300]. He elutes first with $4M$ ammonia, when lead and bismuth remain on the start and other ions move behind the solvent front. Lead and bismuth are then separated in 0.2M sulphuric acid when lead remains on the start and bismuth has R_F 0.88. Separation in the reverse sequence can be achieved by impregnating a zone of the paper with cadmium sulphide. Mercury, silver, copper and bismuth are retained in the sulphide zone, whereas lead passes through [365]. On the other hand impregnating the zone with disodium hydrogen phosphate causes the retention of lead. If radioactive phosphate is used, the excess can be eluted with borate-oxalate buffer, after chromatography with ethanol-5M hydrochloric acid, and the lead can be determined radiometrically [366]. Complexes of lead may also be used for separation, such as the complexes with dithizone [256], 8-hydroxyquinoline [194, 253], and leucine.

Lead can be determined by measuring the area of the spots [367], or polarographically after elution. Photometry can also be employed. The best method is to transform lead diethyldithiocarbamate into the corresponding copper complex [368].

Mercury

In the majority of solvent systems mercury(II) displays maximum mobility. In 1-butanol saturated with 1M hydrochloric acid, mercury moves with the solvent front, while the majority of elements, including silver, lead, copper, iron, etc. remain on or near the start. With increasing hydrochloric acid concentration the chromatographic behaviour of mercury does not change substantially. To separate mercury from certain other elements which are also rather mobile in this solvent (such as cadmium, arsenic and zinc), a mixture of collidine and nitric acid can be used [81]. In this mixture mercury(I) remains on the start, while silver, copper, cadmium, arsenic, antimony, zinc, manganese, cobalt, nickel and certain other elements move at various speeds with the solvent. A mixture of 1-butanol with small amounts of pyridine and water can be used for the separation of mercury(I), silver and lead [81, 83]. These elements can also be well separated in solvents containing mainly chloroform [541]. Thielemann separates mercury from copper and cadmium with 1-butanol saturated with 3M hydrochloric acid [698]. The separation of the two oxidation states of mercury is quite simple. A number of solvent systems are convenient for this purpose, but probably the best is a mixture of t-butyl alcohol with acetone and nitric acid in which mercury(I) remains on the start and mercury(II) has R_F 0.81 [98]. Ivanov and co-workers [369] recommend precipitation chromatography using potassium iodide as the precipitant for the separation of the oxidation states of mercury. Inorganic solvents are also suitable for the separation of copper, mercury, cadmium, lead and bismuth. With 4M ammonia, lead and bismuth remain on the start, and copper, mercury and cadmium move with the solvent front, and can be separated by chromatography at right angles, with a mixture of 0.3M potassium iodide and 0.5N potassium carbonate.

For the separation of mercury, papers impregnated with certain inorganic salts or organic reagents forming complex ions with mercury and certain other elements can be used. Mercury(I), silver and lead can be separated on papers impregnated with sodium chloride or potassium bromide, and mercury(II) and bismuth can be separated on papers impregnated with potassium iodide [370]. Janjić and co-workers [371] used papers impregnated with cadmium sulphide. Of the organic reagents, dithizone [255, 256], cupferron [167], and thioacetamide [257] have been used.

Mercury can be determined quantitatively by planimetry or by comparison of the spot size with spots on a standard chromatogram [84]. Photometric methods are most suitable for the determination after elution from the paper. That using tetraethylthiuram disulphide is best [372]. At higher concentrations of mercury, titration with EDTA may be used.

Selenium and Tellurium

In the separation and determination of selenium and tellurium compounds, chromatography is especially important. The reactions of these two elements are often so similar that many analytical methods will not distinguish between them, and therefore separation is necessary. They can be separated with a mixture of alcohols, e.g. 1-butanol and methanol (6 : 4). In this mixture, tellurium is more mobile than selenium. If a mixture of 1-butanol and 2-propanol (1 : 1) saturated with $7M$ lithium nitrate in $2M$ nitric acid is used, selenium moves faster and both elements are sharply separated from polonium and radium [262]. Good separation not only between selenium and tellurium compounds, but usually also between their oxidation states is possible with a mixture of alcohols and mineral acids. 1-Butanol saturated with $3M$ hydrochloric acid sharply separates tellurium(IV) and tellurium(VI) [261], and a mixture of methanol, ethanol and nitric acid separates distinctly all oxidation states of both elements, and also tellurium(VI) from selenium(VI) [260], the tellurium(VI) remaining on the start, and selenium(VI) moving closely behind the solvent front. Lederer [232] used $2M$ hydrochloric acid in phenol to separate selenium and tellurium, and for their separation from certain accompanying elements. Weatherley [264] recommended mixtures of ethyl methyl ketone with hydrobromic or hydrofluoric acid, or acetone, for the separation of selenium from other accompanying elements in silicates. Selenium moves with the solvent front and it can be detected in quantities as low as 5 μg. For the separation of selenium(IV) and tellurium(IV) by circular chromatography, solvents composed of the usual components are suitable, e.g. acetone, water and acid (7 : 2 : 1). Both mineral and certain organic acids are suitable. The sample should be dissolved in $0.1-0.5M$ sodium hydroxide [542]. For the quantitative separation of selenium from other metal ions use can be made of $0.1M$ ammonium tartrate in $4M$ aqueous ammonia on papers treated with stannic chloride and sodium tungstate [651]. The spot due to selenium moves with the solvent front; other ions (iron, copper, zinc, lead, bismuth and tellurium) remain at the point of application or move only slightly.

For quantitative determination of selenium and tellurium after separation, elution from the paper and photometry is probably most suitable. For the determination of selenium, 3,3,-diaminobenzidine, and for tellurium, bismuthiol II, were found best. Kempe and Denn [543] used the photometric determination of colloidal metallic selenium, after its reduction with potassium iodide, for quantitative determination after separation of selenium(IV) and (VI).

Silver

For the chromatographic separation of silver, solvents containing hydrochloric acid may be used in certain cases. With such solvents silver usually remains on the start in the form of silver chloride, and the proper choice of other solvent components can lead to the required separations. In a mixture of 1-butanol and hydrochloric acid silver is usually separated from mercury and often from lead. This solvent will also separate silver from noble metals, copper and gold. Another group of solvents is based on pyridine or collidine. In these, silver generally has a higher mobility and lead and mercury can also be separated. For the separation of silver, solvents based on acetone are important. Combined with hydrochloric or acetic acid, acetone-containing solvents yield sharp separations from copper, lead, iron, cobalt, nickel, gold and platinum. For the separation of silver from lead, mercury(II) and thallium(II) a mixture of chloroform with more polar solvents, such as alcohols, ketones, esters or acids, may be used [541].

Weiss and Fallab [84] made a semiquantitative determination of copper, silver and mercury by separation using a mixture of 50% ethanol and 50% nitric acid, saturated with benzene. After detection with 0.2% quercetin solution the chromatogram is dried and sprayed with ammonia. The size of the spots is compared with those from a calibration chromatogram. For $15-250$ µg of Ag the error is $20-30\%$. Kielczewski and Tomkowiak [96] determined microamounts of silver planimetrically after separation with a solvent system containing EDTA on paper impregnated with bismuth nitrate and dithizone. Janjić and co-workers [97] determined silver to within 5% by precipitation chromatography on papers impregnated with cadmium sulphide.

Thallium

Thallium can be separated from a number of accompanying elements by using 1-butanol saturated with hydrochloric acid. In 1-butanol saturated with $1M$ hydrochloric acid thallium(III) moves with the solvent front and thus can be separated by paper chromatography [295]. Only gold and mercury(II) move with thallium. In 1-butanol saturated with $3.5M$ hydrochloric acid the mobility of thallium(I) is zero, and it can thus be well separated from mercury, cadmium, arsenic, bismuth, antimony, lead and copper [111]. A sufficiently sharp separation of thallium is possible with a mixture of phenol, methanol and concentrated hydrochloric acid [241]. The R_F value of thallium is about 0.5 in this system. A mixture of methanol, sulphuric acid and water separates thallium from arsenic, mercury, copper and zinc [109], and nitrobenzene saturated with water gives separation of selenium from caesium, potassium and ammonium picrates [294]. Warren and Fink [92] achieved a good separation of a series of elements, including thallium, using a mixture of acetone with various complexing agents. Separations using tartaric acid, disodium hydrogen phosphate, and antipyrine were successful, but tailing occurred if acetone was used without complexing agents. Fişel and co-workers [282] used pentanol saturated with $4M$ hydrochloric acid for the separation of thallium(III). Bhatnagar and Sharma [541] separated thallium(I) from silver, lead and mercury(II) by using solvents containing mainly chloroform.

The most suitable quantitative determination of thallium is probably the photometric determination of its complex with Rhodamine B. The thallium spot is extracted from the paper with $2M$ hydrochloric acid and then determined by reaction with Rhodamine B, extraction with benzene and measurement at 560 nm [282]. If thallium is separated in the form of its picrate [294] the spots can be extracted directly with acetone and the colour intensity compared with that of standard solutions. A semiquantitative determination may be made by comparison of spots detected with 8-quinolinol, with a standard scale [287].

Titanium

The separation of titanium is most commonly required from mixtures containing manganese, iron, chromium, aluminium, cobalt, nickel and zinc or from mixtures with niobium and tantalum. Lacourt and co-workers [288 to 292] thoroughly investigated the separation of mixtures of iron, aluminium and titanium, and achieved good results both with separation and quantitative determination. They succeeded in sharply separating the series (in descending order of R_F) Fe > Ti > Al by using a mixture of pentanol, benzene and hydrochloric acid. 1-Butanol mixed with hydrochloric acid [155] separates titanium, which remains near the start, from elements which are more mobile in this mixture. The R_F value of titanium in 1-butanol and nitric acid is substantially increased if they are mixed with some acetylacetone and thus it may be separated from vanadium, iron and molybdenum [212]. Acetylacetone mixed with 2-propanol and hydrochloric acid can be used for the separation of titanium, beryllium and uranium after their separation from the majority of other ions by precipitation with ammonia from a medium containing EDTA [143]. Ethanol and hydrochloric acid [265] also enable a number of separations to be made, especially from thorium, zirconium and molybdenum. Distinct separation from niobium, tantalum, uranium, zirconium and hafnium can be obtained with 1-butanol mixed with hydrochloric and hydrofluoric acid [126]. Martin and Magee [131] used diethyl ketone saturated with $2M$ hydrofluoric acid and $2M$ nitric acid to separate niobium, tantalum and titanium. For the same separation Cabral [138] used ethyl methyl ketone and hydrochloric acid. Qureshi and Khan separate titanium from 32 other species by developing with formic acid, hydrochloric acid and acetone (3 : 5 : 2) [652]. Titanium(III) and (IV) can be separated with methanol containing ascorbic acid or acetoacetic ester, and also with a mixture of acetone and ascorbic acid [330]. Elbeih and Gabra [331] made use of EDTA complexes for the determination of iron and titanium in ilmenites. They used a mixture of EDTA in $1M$ nitric acid, 1-butanol and acetylacetone.

One method of quantitative determination is to use detection with kojic acid and hydroxyquinoline and to compare the sharply defined spots with spots on calibration chromatograms. Photometric methods, however, either direct [290] or after elution, are better. For accurate determination the spot should be cut out and ashed. The resultant oxide is dissolved in a mixture of hydrofluoric and nitric acid, and, after addition of sulphuric acid, evaporated to dryness. The residue is dissolved in hydrochloric acid, and the titanium is then determined by a suitable method [131].

Thorium

Thorium usually has to be separated from mixtures of uranium, thorium, rare earths and zirconium. Most chromatographic methods published for thorium deal with this separation. Generally uranium has a higher R_F value than thorium. In diethyl ether — nitric acid, thorium remains on the start, while uranium moves sufficiently [312]. The same happens with a mixture of acetone, isobutyl alcohol and hydrochloric acid [272]. Thorium may be separated from rare earths and zirconium by using a mixture of tetrahydrosilvan, nitric acid and water, or a mixture of methyl acetate, nitric acid and water [313]. A mixture of dioxan and antipyrine does not separate thorium from uranium sharply, but does give a distinct separation of thorium from the majority of other accompanying elements [315]. Zirconium can be separated from thorium by mixtures of ethanol and hydrochloric acid, in which the mobility of thorium is greater than that of zirconium which remains near the start [265]. On the other hand, a mixture of isobutyl alcohol and pyridine separates these elements so that zirconium moves faster than thorium [267]. Thorium can be separated from zirconium and uranium with a mixture of methyl isobutyl ketone, isobutyl alcohol and tributyl phosphate (25 : 19 : 6) [547]. Thorium, uranium, and rare earths can also be separated by reversed-phase chromatography. Tributyl phosphate is suitable as the stationary phase, and a solution of $0.2M$ ammonium thiocyanate in $0.1M$ nitric acid is a suitable mobile phase [545]. Gupta and Mukerjee [544] impregnated the paper with glycerol and used phenol saturated with water as the mobile phase.

For quantitative determination both photometric methods (e.g. the reaction with anthranilic acid) and titrimetric methods [315] can be used. The thorium spot, after detection with a mixture of kojic acid and oxine, is eluted with water slightly acidified with nitric acid and added to a known amount of EDTA. A known and excessive volume of bismuth nitrate solution is added and the solution back-titrated with $0.001M$ EDTA. Născuţiu recommends irradiation of the chromatogram with a beam of thermal neutrons and subsequent scanning of the radioactivity for the determination of thorium as well as a number of other elements [548].

Uranium

Uranium was one of the first elements to be studied chromatographically. This is due to its ability to form a carbonate complex readily and its high mobility in various mixtures of organic solvents with nitric acid. Arden and co-workers [323] proposed a mixture of tetrahydrofuran or tetrahydropyran with nitric acid as the most suitable solvent and were able to perform very sharp separations, so that the method became a basis for the determination of uranium in minerals and ores. A mixture of ether and methanol with nitric acid has been used for the sharp separation of uranium(VI) and iron [322]. A similar separation from copper can be done by using a mixture of ethanol and sulphuric acid [240]. Very good separations can be obtained by using simple inorganic solvents, such as a solution of ammonium carbonate and ammonium bicarbonate, as uranium readily forms carbonate complexes. In this solvent uranium moves with the solvent front, while the majority of other elements have R_F values lower than 0.76 [321]. The carbonate complex can be used in a slightly different manner, by forming the complex first, followed by separation (e.g. with a mixture of ethyl methyl ketone, acetyl-acetone and acetic acid) [324]. Sometimes it is more convenient to concentrate the uranium by a suitable method before the separation, such as precipi-tation with ammonia in a solution of EDTA. The precipitated uranium, beryl-lium and titanium may be sharply separated with a mixture of 2-propanol, acetylacetone and hydrochloric acid [148]. Mixtures of butanol with hydro-chloric acid of various concentrations also give such separations, and in addition, separation of uranium(VI) from uranium(IV) and separations from mixtures containing transuranium elements [320]. The separation of uranium from thorium has been extensively studied. The following solvents will separate them: alcohol, dioxan and pyridine [311], dioxan, hydrochloric acid and antipyrine [315], methyl isobutyl ketone, isobutyl alcohol and tri-butyl phosphate (25 : 15 : 6) [547] or simply butanol and hydrochloric acid [316]. Good results are also obtained with reversed-phase chromatography, e.g. with tributyl phosphate as the stationary phase and $0.2M$ ammonium thiocyanate in $0.1M$ nitric acid as the mobile phase. Complexes with 1-nitro-so-2-hydroxy-3-naphthoic acid can be used for the separation of uranium from thorium, zirconium, cobalt and palladium, a suitable solvent being acetone, 2-butanol and hydrochloric acid [272]. Nagai and Deguchi [167] separated iron, copper, zinc, mercury and uranium by circular precipitation chromatography based on the separation of cupferronates of these metals.

For quantitative determination of uranium after separation a number of methods have been proposed. After detection with ferrocyanide the well-defined spots can be compared with a series of standards or measured planimetrically or photometrically, However, the most common method is elution of the separated uranium from the paper and polarographic, colori-metric or fluorometric determination [323]. The chromatograms can also be evaluated by autoradiographic methods [327].

Vanadium

Vanadium has a low mobility in solvents containing lower aliphatic alco-hols. Mixtures of 1-butanol and hydrochloric acid will separate vanadium from elements which are more mobile in this solvent such as cadmium, iron, bismuth, mercury and zinc. Increasing the concentration of the hydrochloric acid does not increase the mobility of vanadium substantially. This is also the case in solvents containing pentanol and hydrochloric acid. In solvents containing lower alcohols, especially methanol, and acids, the mobility increases, but the separating power of these mixtures is not very good. Vanadium has its highest R_F values in solvents containing mainly acetone.

The separation of vanadium from titanium, niobium and tantalum is most often required. Solvents containing acetone, hydrochloric acid and oxalic acid [87] are best for this. The metals in the form of their oxalate complexes move in the order Fe > Nb > V > Ta > Ti. The separation of vanadium from molybdenum or tungsten is also a frequent task. All three may be sharply separated with a mixture of pentanol and formic acid [306]. The separation may also be done with a mixture of 1-butanol, anisole and hydrochloric acid, or pentanol, dioxan and hydrochloric acid. The best solvent for the separation of vanadium from uranium is tetrahydrofuran and ether, but an adequate separation can also be achieved with mixtures of various ketones and mineral acids. The separation of vanadium from a series of other ions by mixtures of diethyl ether or tetrahydrosilvan with nitric acid and hydrogen peroxide has been described by Arden and co-workers [280]. Stevens [307] separated some oxidation states of vanadium by using a mixture of ethanol or acetone with acetic acid and sodium acetate. Vanadium can be separated from large amounts of lead, cadmium, chromium, magnesium, zinc, aluminium, manganese, copper, molybdenum, bismuth and tungsten in the form of oxinates by eluting with a mixture of isopentyl alcohol and acetic acid [308].

8-Hydroxyquinoline is also the best means for quantitative photometric determination of vanadium after separation [309]. Kotsuji recommended chromatographing with dilute sulphuric acid alone, when vanadium(IV) forms a blue zone which after elution can be determined photometrically with oxine at 550 nm.

Bismuth

For the separation of bismuth from other ions solvents containing mainly acetone are the most convenient [92]. In such solvents bismuth usually moves with the solvent front, while the mobility of other elements is lower. In mixtures of acetone with hydrochloric acid sharp separation of bismuth from lead, copper, nickel, cobalt and thallium can be achieved. The separation can be further improved by the addition of complexing agents to the acetone. Separation is sharpest with a mixture of acetone and methyl acetate. While bismuth moves together with gold, platinum and cadmium at the solvent front, the majority of other elements remain on the start. In mixtures of 1-butanol with hydrochloric acid bismuth has a medium R_F value. Therefore, bismuth can be separated quite well from those ions which have low R_F values in these solvents, such as nickel, cobalt, aluminium, silver and copper, and also from ions which move substantially faster, such as iron. Separation in inorganic solvents, proposed by Imai [300], is an interesting possibility. When a mixture of copper, mercury, cadmium, lead and bismuth is separated in $4M$ ammonia the last two elements remain on the start, while the R_F values of the first three are in the range $0.85-1.0$. Lead and bismuth can be separated in $0.2M$ sulphuric acid, lead remaining on the start, and bismuth moving with R_F value 0.88. In a mixture of ethyl methyl ketone and hydrofluoric acid [263], bismuth remains on the start and can be clearly separated from selenium, tellurium and polonium. This is an important separation in the determination of polonium in preparations containing bismuth in analysis of radioactive mixtures. Okáč and Černý [302] demonstrated that bismuth, antimony and tin can be separated from an acid solution with acetate and then by capillary chromatography in ammonium polysulphide. Complexes with some organic reagents can also be used for separation, e.g. the separation of dithizonates [256], and also the separation of mercury(II), copper(II), bismuth and cadmium ions in the form of their sulphides, with thioacetamide [257].

For quantitative determination of bismuth in ores Demetrescu and Cioaca [146] eluted the spots from the chromatograms and used photometry in the presence of thiourea. For a similar analysis Agrinier [301] used photometry with 2,5-dimercapto-1, 3, 4-thiodiazole. Sikorska-Tomicka [303] used a solution of caprolactam and iodine in acetone for the detection at 420 nm after eluting the spot with acetone.

Tungsten

Tungsten does not lend itself to chromatographic separation and only a few papers deal with it. It is most commonly separated from mixtures with molybdenum, vanadium and sulphate, with a mixture of pentanol and formic acid [306] or solvents containing various ketones and various amounts of hydrochloric acid [318]. In both cases tungsten remains on the start. It can be separated from vanadium and chromium with an isobutyl alcohol, collidine, hydrochloric acid mixture [427]. Tungsten can be separated from a series of other elements, such as iron, molybdenum, niobium and titanium, which move with the solvent front, by a mixture of ethanol, methanol and $2M$ hydrochloric acid, or of acetone with hydrochloric and hydrofluoric acids.

Maleczki [319] describes a semiquantitative determination of tungsten, after a separation using ethanol, ammonia, water and sodium potassium tartrate (40 : 5 : 55 : 2), by detection with a solution of vanadyl nitrofluorone in sulphuric acid, and comparison of the spots with those on standard chromatograms.

Zinc

The chromatographic behaviour of zinc has been widely studied. A number of solvent systems have been proposed for sharp separations from various mixtures. Thielemann has recently used a dioxan — nitric acid — water mixture (200 : 2 : 5) to separate zinc from other metal ions found in effluents from plating plant [698]. 1-Butanol mixed in various proportions with hydrochloric acid usually produces sufficiently sharp separation from copper, lead, bismuth, mercury, aluminium, nickel and cobalt. A 1-butanol, ammonia and pyridine mixture produces sharp separation of zinc from cadmium [161]. Zinc can also be separated from iron in solvent systems containing mainly 1-butanol. A large number of good separations have been achieved with acetone solvent systems. Thus, zinc which usually moves with the solvent front, can be separated from iron, aluminium, chromium, manganese, nickel, cobalt, molybdenum, uranium, vanadium and titanium. In solvents based on acetone, containing acetylacetone, isobutyl alcohol and hydrochloric acid, the separation sequence is changed [162]. Zinc moves closely behind the solvent front, but iron has a lower mobility and together with copper and lead can be separated from zinc. The separation of pyridine complexes by a mixture of ethanol and acetone is also of interest. Under these conditions zinc can be well separated from cadmium, nickel, cobalt and copper. For the separation of zinc from copper, chromatography of the oxinates with 1-butanol and hydrochloric acid has been proposed [158]. Zinc is well separated from arsenic and thallium [109] in a mixture of methanol and sulphuric acid. Pickering [166] used paper chromatography in a study of the cyanide complexes of numerous metals, and he demonstrated that various separations of zinc are possible if suitable conditions are chosen. Qureshi examined the EDTA complex [489]. Venturello and Ghe [165] investigated separation of the dithizonates of zinc, mercury, silver, copper, cadmium, lead and bismuth on papers impregnated with buffers of various pH values $(1-10)$, by solvents based on certain aliphatic alcohols. They found that the dithizonates of metal in the same group of the periodic table displayed similar chromatographic behaviour. Nagai and Deguchi [535] separated cobalt and zinc by first impregnating the paper with dithizone and chromatographing with a mixture of $3M$ acetic acid and acetone $(1:1)$. The chromatographic separation of metal cupferron complexes can be very efficient [167]. Distinct separation of mixtures such as iron, copper, mercury and zinc, or iron, copper, zinc and uranium, can be achieved on circular papers impregnated with an alcoholic solution of cupferron, by using 16% acetic acid as solvent. Distinct separation of zinc from a series of other ions by 1-butanol−concentrated hydrochloric acid is obtained by circular chromatography [168]. In Imai's method of separation on chromatographic paper impregnated with 1-nitroso-2-naphthol at R_F $0.1-0.2$ and dimethylglyoxime at R_F $0.4-0.5$ (cf. p. 84), with an inorganic solvent, such as ammonium chloride, thiocyanate, nitrate or sulphate, manganese remains on the start, nickel and cobalt are retained in the impregnated zones, and zinc moves up to the zone of R_F $0.55-1.0$, where it can be determined after detection with 8-hydroxyquinoline and observation under ultraviolet light.

For quantitative determination of zinc the sizes of the detected spots can be compared visually [170, 171, 176]. Photometric and titrimetric methods can also be used. Lacourt and co-workers devoted great attention [172, 177, 173] to the last two methods. They investigated the determination of microamounts of the elements separated and proposed a direct titration with a solution of EDTA and photometric methods based on the reaction with dithizone and diphenylcarbazide, either after elution from the paper or directly on the paper chromatogram. Dithizone has also been recommended by other authors as a suitable reagent for photometric determination of zinc after chromatography [159, 163, 174]. Frierson and co-workers [175] proposed Zincon (2-carboxy-2′-hydroxy-5′-sulphoformazylbenzene) as a suitable reagent for the determination of zinc in a mixture of cobalt, nickel and zinc, because with zinc it gives a strong colour with an absorption maximum at 620 nm.

Zirconium and Hafnium

Zirconium and hafnium are chemically very similar indeed. They are separated only with great difficulty by ordinary chemical or physical methods, and extraction or chromatographic methods are generally used for this purpose. Good results have been achieved with paper chromatography in solvents based on tributyl phosphate, 1-butanol, and xylene or benzene, saturated with $8-10M$ nitric acid. Under optimum conditions the R_F values of the two elements differ by as much as 0.5 [268]. Crawley [269] showed that partial separation of zirconium and hafnium can be achieved with tributyl phosphate acidified with nitric acid, and that the addition of sodium nitrate improves the separation. For preparative purposes zirconium and hafnium may be chromatographed on a silica gel column with methanol [270]. The hafnium content of zirconium can thus be reduced to below 0.1%.

Special attention is often devoted to the separation of zirconium from titanium, thorium, niobium, molybdenum, uranium, and other accompanying elements. These separations can usually be achieved successfully with mixtures of ethanol and hydrochloric acid, sometimes with the addition of water [265]. Sharp separation of zirconium from niobium and molybdenum can be achieved with ethyl methyl ketone and dioxan from a medium containing oxalic and hydrochloric acids [139]. Zirconium and thorium can be separated in isobutyl alcohol, water and pyridine, in the presence of thiocyanate [267]. Good separation of zirconium (R_F 0.0) from thorium, uranium, copper, iron and cerium (R_F 0.8 − 1.00) can be obtained with a mixture of hydrochloric acid, phosphoric acid and water (10 : 1 : 9) [548]. Datta and Ghose separated cobalt, palladium, uranium, thorium and zirconium from other elements by precipitation with 1-nitroso-2-hydroxy-3-naphthoic acid followed by chromatography of the precipitated complexes with various solvents in which zirconium and thorium remain on the start [272]. The separation can also be done with a mixture of methyl isobutyl ketone, isobutyl alcohol and tributyl phosphate [547]. According to Născuţiu and Grigorovici small amounts of hafnium and niobium can be separated from large amounts of tantalum with a mixture of acetylacetone or dioxan, hydrofluoric acid, and $0.5M$ oxalic acid [539]. Zirconium and niobium in hydrofluoric acid can be separated by a 4 : 1 mixture of 1-butanol with $7M$ lithium nitrate/$2M$ nitric acid [540].

The application of chromatographic methods to quantitative determination of zirconium and hafnium was developed by Elbeih and Gabra [271]. In analysis of ores, after decomposition, zirconium and hafnium were separated from accompanying elements with a mixture of phenazone, dioxan and nitric acid, and the sum of both elements determined.

Gold

Gold is most often to be separated from other elements of the platinum group. The separation within this group is difficult and only a few solvent systems are effective. There are two groups of solvents which are generally suitable. The first is based on 1-butanol and hydrochloric acid and usually gives a sharp separation of gold from other platinum metals, as well as a separation within the latter. Thus, for example, 1-butanol saturated with $1M$ hydrochloric acid separates the metals in the following order of decreasing R_F: Au > Pt = Ir(IV) > Pd > Ru = Rh [374]. The second group consists of ketones such as ethyl methyl ketone, methyl propyl ketone, isopropyl methyl ketone, or methyl isobutyl ketone in combination with hydrochloric acid. These solvents also separate gold well from the platinum metals. Gold usually moves with the solvent front or closely behind it and the other metals in the sequence (of decreasing R_F) Os > Pt > Pd > Rh = Ir(III) = Ru. Gold is most sharply separated from the platinum metals by a mixture of ether, methanol and hydrochloric acid [120] in which gold moves with the solvent front and platinum metals remain on the start. A similar sharp separation is possible with a mixture of ethyl acetate, water and nitric acid [375]. The separation of gold from platinum metals in inorganic solvents is both interesting and effective. In $0.1-6.0M$ ammonium chloride, $0.1-5.0$ sodium chloride, and $4-8M$ hydrochloric acid [373] the mobility is usually reversed, e.g. $4M$ hydrochloric acid separates the metals as follows, iridium (R_F 0.98), rhodium (0.87), ruthenium (0.84), palladium (0.75), platinum (0.67), osmium (0.56) and gold (0.22). The separation of gold and platinum metals is also possible with a mixture of acetophenone, acetone and hydrochloric acid (5 : 2 : 3) [549].

For quantitative determination of gold, elution from the paper is usually necessary. Anderson and Lederer [376] determined gold electrolytically, but photometric methods are more common, e.g. formation of the orange complex bromoaurate(III) ion, or determination with Rhodamine B.

Iron

The mobility of iron(III) in solvents consisting of 1-butanol and hydrochloric acid depends on the acid concentration. In a mixture of 1-butanol and $1M$ hydrochloric acid iron remains near the start, but in 1-butanol saturated with $3M$ acid the R_F is $0.5-0.6$. The mobility increases steadily up to 1.0 with increasing concentration of the acid. This is due to the formation of chloro-complexes which are more mobile in organic solvents. A high mobility is also observed in solvents containing acetone and hydrochloric acid, in which iron moves with the solvent front and thus is separated from copper, cobalt and nickel which move more slowly and in that order. A mixture of acetone, isobutanol, acetylacetone and hydrochloric acid separates iron from nickel, copper and lead [162]. Separation from aluminium is simple. In mixtures containing 1-butanol and high concentrations of hydrochloric acid, aluminium remains on the start, while iron, as already said, moves with the solvent front. Ethyl methyl ketone and hydrochloric acid will separate iron, beryllium and aluminium [151]. The separation of iron, titanium and aluminium was investigated in detail by Lacourt and co-workers [288−290, 292]. Using various solvents, especially a mixture of pentanol, hydrochloric acid and benzene, they were able to sharply separate iron, titanium and aluminium (in order of decreasing R_F) and to develop procedures for quantitative determinations.

It is often important to separate quantitatively both oxidation states of iron. This can easily be done by chromatography. The simplest solvents are

$4M$ hydrochloric acid and 1-butanol, or $4M$ hydrochloric acid, 1-butanol and acetone [336].

The separation of complex ions of iron has also been thoroughly investigated. Barbieri and co-workers [337] separated phosphate complexes of iron, and Hill [338] studied the chromatographic separation of a large variety of iron complexes, e.g. with ethylenediamine, iminodiacetic acid, glycine and EDTA. The EDTA complex has also been investigated by Qureshi [489]. Elbeih and Abou-Elnaga recommended the use of solvents containing benzoylacetone for the separation of iron [550]. A reliable separation of iron from manganese, nickel and cobalt can be obtained with methylamine [534]. Japanese workers investigated separations on papers impregnated with organic reagents. Nagai [254, 339, 340] developed a method of precipitation chromatography using 8-hydroxyquinoline and in particular was able to separate copper and iron. Yokota and Shimada [258] chromatographed iron, copper, manganese, and nickel on paper impregnated with kojic acid. Nagai and Deguchi [167] developed a method of circular chromatography, separating iron, copper, zinc and uranium in the form of their complexes with cupferron. Ćelap and co-workers [341] used paper impregnated with nickel sulphate and potassium ferrocyanide for the separation of lead, zinc, cadmium and iron.

For quantitative determination of iron after its chromatographic separation it should be eluted from the paper. In the eluate it can be determined simply by photometry, for example with Tiron at pH 4.7 [289], or Morin at pH 5.0 [550]. However, it is possible to use other colorimetric reagents, for example thiocyanate or thioglycollic acid. Among direct methods, planimetric evaluation of the radiograph of a chromatogram containing ^{59}Fe-labelled material may be mentioned [529, 530].

Platinum Group Metals

Chromatographic separation of the metals of the platinum group has already been mentioned (see p. 108). In addition to the solvents containing mainly 1-butanol or ketones and hydrochloric acid described in the section on gold, mention should also be made of some other possible solvents. With an ethyl acetate, hydrochloric acid and acetic acid mixture, platinum, osmium and palladium can be separated from ruthenium, rhodium and iridium, which remain unseparated near the start [381]. A mixture of methyl isobutyl ketone and hydrochloric acid [383] separates the last three metals. Lederer [77] proposed solvents containing stannous chloride for separation of the platinum metals. The different behaviour of tin(II) with platinum metals on chromatograms permits separation, with $0.5N$ tin solution in $3M$ hydrobromic acid, of ruthenium, osmium, palladium, rhodium and platinum (in order of decreasing R_F). Iridium forms three distinct spots, coinciding with platinum, rhodium and ruthenium. An explanation of this has been given by Bagliano [379] in a study of the complexes of the iridium(II) and (IV) with stannous bromide. A clear separation of rhodium, palladium and platinum from one another as well as their separation from gold, is possible with a mixture of methyl isobutyl ketone and hydrochloric acid [383]. An adequate separation of palladium, ruthenium, indium, rhodium, platinum

Table 6.3. Review of Some Separations of Platinum Metals

Solvent system	R_F	Reference
Ethyl acetate + HCl + AcOH (60 : 30 : 10)	Ru 0.10; Rh 0.18; Pd 0.42; Os 0.53; Ir 0.19; Pt 0.58	[377] [381]
Methyl isobutyl ketone + 10M HCl (1 : 1)	Ru 0.25; Rh 0.34; Pd 0.52; Os 0.87; Ir 0.0; Ir 0.35 (two spots); Pt(II) 0.52; Pt(IV) 0.76; Cu 0.63; Au 0.99	[382]
1M SnCl$_2$ in 6M HCl	Pt 0.21; Pd 0.70; Rh 0.13—0.56; Os 0.69; Ru 0.77; Ir 0.18; 0.78 (two spots)	[77]
0.5M Sn in 3M HBr	Pt 0.24; Pd 0.6; Rh 0.37; Os 0.76; Ru 0.81; Ir 0.19; 0.32; 0.82 (three spots)	[77]
Methyl isobutyl ketone + HCl (70 : 30)	Rh 0.14; Pd 0.41; Pt 0.68; Au 0.97	[383]
Acetone + iBuOH + 12M HCl (60 : 38 : 2)	Pd 0.29; Co 0.45; U 0.53 Th 0.0; Zr 0.0	[272]

and gold can be achieved with a mixture of isobutyl alcohol, ethanol, water and acetic acid (60 : 57 : 80 : 3) [552]. Day and Chakrabartty [551] systematically investigated 48 different solvent systems based on mixtures of lower aliphatic alcohols and acetone, or dioxan with water and hydrochloric acid, containing a chelating, oxidizing, or reducing agent. They concluded that perfect separation of all ions of the platinum group is scarcely possible, but that good separation of up to five ions of this group may be obtained.

Paper chromatography permits study of the reactions of complex compounds. The chloro, ammino, pyridino, and chloropyridino complexes of platinum(II) [384] can be clearly separated chromatographically and so can the *cis* and *trans* dihalodiammino complexes [385]. Chloride complexes of rhodium [380], and iodide and chloride complexes formed on reaction of osmium tetroxide with potassium iodide in hydrochloric acid may also be separated [378].

Photometric methods are usually most suitable for quantitative work. For platinum and rhodium, reaction with stannous chloride, and for palladium, reaction with thioglycollic acid or *p*-nitrosodimethylaniline, are convenient. For the determination of iridium, redox titration with hydroquinone [383] has been recommended as most suitable.

A review of some separations of platinum metals is given in Table 6.3.

Rare Earths

The separation and determination of the rare earths, which form a homogeneous group of 14 elements, is a difficult problem in analytical chemistry.

At present paper chromatography gives neither an absolute and simultaneous separation of all members of this group, nor a separation of rare earth elements from other substances, even though the chromatographic behaviour of elements of this group has been thoroughly investigated. The separation of the uncomplexed ions on pure chromatographic paper does not appear to be possible for the whole group, although elements at one end of the group can be separated from those at the other. The problem was studied in great detail by Lederer [393–395], who demonstrated that with solvents based on lower alcohols and acetone with hydrochloric acid the R_F values were dependent on the solvent composition. In such mixtures the R_F values usually increase with atomic number, but the difference in R_F value between elements close to each other in the series is usually so small that it does not permit a separation. For this type of separation, butanol, acetylacetone and acetic acid is probably the most suitable solvent mixture [388], because the R_F values of rare earth elements range from about 0.2 to 0.6. Thus, this system permits separation of a series of groups of rare earth elements.

The possibility of separating these elements by paper chromatography is increased by using impregnated papers. Organic reagents usually do not increase the sharpness of separation but only increase the reliability of the separation of some lanthanides in qualitative analysis, e.g. papers impregnated with 8-hydroxyquinoline [396]. Sometimes it is more convenient not to treat the paper with organic agents, but to add these to the solvent system, as demonstrated by Pollard and co-workers [392] in the case of 8-hydroxyquinoline and a mixture of 1-butanol and acetic acid.

The sharpest separation of this series can be made on papers impregnated with inorganic salts, amongst which silver nitrate is most effective. Poluektov and Lauer [398, 399] successfully separated single elements in order of their atomic number by using a solvent system containing acetone, ether, water, ammonium thiocyanate and hydrochloric acid. In an ascending arrangement combined with over-running [400] lanthanum, cerium, praseodymium, neodymium, samarium, gadolinium, yttrium, holmium and ytterbium can be sharply separated.

Chromatographic methods can also be used for preparative purposes. Kiba and co-workers [180] separated yttrium from strontium, and were able to prepare ^{90}Y in 99.9% radiochemical purity by using a mixture of ethanol and aqueous ammonium thiocyanate solution. Weil [401] applied the methods of industrial chromatography to the separation of the rare earth elements.

One method of quantitative determination of separated rare earths is to detect the separated zones with a solution of alizarin in ethanol, then cut them out, ash them, dissolve the residues in nitric acid, evaporate, extract with water and determine the element photometrically after reaction with sodium alizarin sulphonate or arsenazo I [397]. Năşcuţiu [548] recommended irradiation of the chromatogram with a stream of thermal neutrons and determining the activity of individual spots: the relative error was 12%. For the determination of scandium after separation from uranium, lanthanum, gadolinium and ytterbium with a mixture of ether, acetone, $0.82M$ ammonium thiocyanate and concentrated hydrochloric acid (150 : 150 : 4 : 3), titration with $0.0001M$ EDTA, using 0.1% Xylenol Orange solution as indiatorc, has been suggested [553]. EDTA can also be employed for quantitative

Table 6.4. Review of Some Separations of Rare Earths (R_F Values Given)

Element	Solvent							
	1	2	3	4	5	6	7*	8*
Y	0.70	0.58	—	0.48	0.59	0.03	—	—
La	0.71	0.40	—	0.08	0.31	0.95	—	0.23
Ce	0.65	0.46	—	0.11	0.38	0.92	0.27	—
Pr	—	0.51	—	0.16	0.38	0.92	—	—
Nd	—	0.54	—	0.20	0.22	0.90	0.40	—
Pm	—	—	—	—	—	—	—	—
Sm	—	0.55	—	0.31	0.47	0.84	—	0.40
Eu	—	0.55	—	—	0.49	0.74	—	—
Gd	—	0.56	—	0.44	0.48	0.60	0.54	0.53
Tb	—	—	—	—	—	0.25	—	—
Dy	—	0.62	—	0.50	0.62	0.10	—	—
Ho	—	—	—	—	0.56	0.05	—	—
Er	—	0.64	—	0.56	0.60	0.03	—	—
Tm	—	—	—	—	—	0.0	—	—
Yb	—	—	—	0.59	0.59	0.0	—	—
Lu	—	—	—	—	—	0.0	—	—
Sc	—	1.0	0.57	—	0.97	0.0	—	—
Sr	0.32	—	—	—	—	—	—	—
Ba	0.40	—	—	—	—	—	—	—
Re	—	—	0.39	—	—	—	—	—
Th	—	—	0.77	0.90	—	—	—	—
Ti	—	—	0.38	—	—	—	—	—
Zr	—	—	0.30	—	—	—	—	—
Ac	—	—	—	—	0.68	—	—	—
U	—	—	—	—	—	—	—	—
Ref.		[386]	[313]	[387]	[388]	[389]	[390]	[390]

Solvent compositions:

1 — MeOH + 6M HCl + 5% citric acid (1 : 1 : 8).
2 — BuOH saturated with 7M LiNO$_3$/2M HNO$_3$.
3 — MeOAc + HNO$_3$ + H$_2$O (85 : 5 : 10).
4 — Et$_2$O + Me$_2$CO + HSCN.
5 — BuOH + Me$_2$CO + AcOH + H$_2$O (100 : 30 : 5 : 65).
6 — Na$_2$SO$_4$ solution (acidified); paper impregnated with di(2-ethylhexyl) phosphate in benzene.
7 — 80% MeOH + 2.4M HNO$_3$; Whatman DE-20 paper.
8 — 90% MeOH + 0.6M HNO$_3$; Whatman DE-20 paper.
* R_F depends on distance travelled by solvent front.

Table 6.4 continued

Element	Solvent								
	9*	10*	11*	12*	13*	14	15	16	17
Y	—	—	—	—	—	—	0.07	0.47	0.16
La	—	—	—	0.06	0.17	0.87	0.92	0.34	0.11
Ce	0.13	0.24	0.06	—	0.11	—	0.91	0.40	—
Pr	—	0.38	0.12	—	—	—	0.90	0.40	—
Nd	0.29	—	0.21	0.33	0.37	—	—	0.43	—
Pm	—	—	—	—	—	—	—	—	—
Sm	—	—	0.40	0.56	0.60	—	0.76	0.44	—
Eu	0.46	—	—	—	—	—	0.64	—	—
Gd	—	—	0.60	—	0.61	—	0.51	0.44	—
Tb	—	—	—	—	—	—	—	0.47	—
Dy	—	—	—	—	—	—	0.12	0.50	0.20
Ho	—	—	—	—	—	—	—	—	—
Er	—	—	—	—	—	—	0.06	0.50	—
Tm	—	—	—	—	—	—	—	0.66	—
Yb	—	—	—	—	—	—	0.03	0.70	0.17
Lu	—	—	—	—	—	—	—	0.59—0.76	—
Sc	—	—	—	—	—	—	0.0	—	—
Sr	—	—	—	—	—	—	—	—	—
Ba	—	—	—	—	—	—	—	—	—
Re	—	—	—	—	—	—	—	—	—
Th	—	—	—	—	—	0.10	0.0	—	—
Ti	—	—	—	—	—	—	—	—	—
Zr	—	—	—	—	—	—	—	—	—
Ac	—	—	—	—	—	—	—	—	—
U	—	—	—	—	—	0.51	—	—	—
Ref.	[390]	[390]	[390]	[390]	[390]	[317]	[391]	[392]	[393]

Solvent compositions:

 9 — 90% MeOH + 1.2M HNO$_3$; Whatman DE-20 paper.
10 — 95% MeOH + 0.15M HNO$_3$; Whatman DE-20 paper.
11 — 99% MeOH + 0.15M HNO$_3$; Whatman DE-20 paper.
12 — 99.66% MeOH + 0.05M HNO$_3$; Whatman DE-20 paper.
13 — 99.9% MeOH + 0.15M HNO$_3$; Whatman DE-20 paper.
14 — 2M NH$_4$NO$_3$; paper impregnated with tri-*n*-octylamine.
15 — 1M HCl; paper impregnated with 0.1M di(2-ethylhexyl) phosphoric acid.
16 — BuOH + oxine + AcOH (100 ml : 5 g : 20 ml).
17 — EtOH + 2M HCl (9 : 1).
* R_F depends on the distance travelled by the solvent front.

determination of other elements of this group. After detection on the chromatogram with arsenazo I, they were extracted and determined by back-titration with lanthanum after the addition of excess of EDTA [554].

A review of some separations of rare earths is given in Table 6.4.

Technetium, Rhenium

Derivatives of rhenium and technetium can be separated by chromatography of the products formed on heating with $2M$ nitric acid in the presence of excess of thiourea [76], $2M$ hydrochloric acid being used as solvent. Satisfactory separation is also given by using ethanol or butanol saturated with hydrogen iodide, on paper impregnated with hydrazine sulphate [402]. Butanol mixed with water and perchloric acid may also be used [555]. The behaviour of technetium and rhenium during paper chromatography and ion-exchange chromatography was studied by Ossicini and co-workers [403]. They mainly studied the behaviour of compounds of rhenium(VII) and technetium(VII) on various ion-exchange papers and followed the reactions of technetium derivatives with hydrochloric and hydrobromic acids by chromatography at room temperature and at 100°. Matsuura and Kojima [404, 405] investigated chromatographically the relationship between the R_F values of pertechnic acid ($HTcO_4$) and its partition coefficient and investigated the existence of compounds of technetium(IV) in hydrochloric acid. The paper chromatographic separation of perrhenate in solvent systems containing pyridine or its alkyl derivatives, 1-propanol, 2-propanol or ethyl methyl ketone was studied by Toul and Tusl [653]. The best separation was achieved by using pyridine — $13M$ ammonia (10 : 1). Perrhenate has an R_F value of 0.98; many other ions remained at the origin. Partial migration of several other ions was decreased by preliminary treatment of the sample solution with alkali.

Polonium

Hydrochloric or hydrobromic acid solutions can be used to separate a number of heavy metals by paper chromatography, as chloro- or bromo-complexes. In addition to polonium, gold, rhenium and technetium belong to this group. Polonium can be separated from a number of other elements on acetylated cellulose paper with a mixture of 1-butanol and hydrochloric acid as solvent [406]. However, if a nitric acid solution of polonium is applied to the start, two distinctly separated spots are formed [407].

Separation of polonium from selenium and tellurium, and also from radium, is the most often required. Both separations can be achieved by using a 1-butanol/1-propanol (1 : 1) mixture saturated with a mixture of $7M$ lithium nitrate and $2M$ nitric acid [262].The following R_F values have been measured: RaD 0.27, Te 0.47, RaE 0.62, Se 0.72 and Po 0.97.

Polonium can be separated from tellurium, selenium and bismuth with a solvent composed of 60 g of 49% hydrofluoric acid in 100 ml of ethyl methyl ketone. The following R_F values were obtained: Se 1.0, Ta 0.63, Po 0.51, Bi 0.0 [263]. Lead, bismuth, and polonium can be separated with a mixture of water and butanol saturated with $3M$ hydrochloric acid. Polonium was separated from tellurium with butanol saturated with $3M$ hydrochloric acid or with a mixture of 2-propanol, trichloroacetic acid and water [654]. Tauber and Schönfeld [408] demonstrated that polonium can be adsorbed on cellulose and on glass equally well, and investigated the influence of pH and acid concentration on the process, which they concluded was adsorption of radiocolloidal aggregates.

Francium, Actinium

The separation of actinium and francium has been investigated by Perey and Adloff [474, 475]. They demonstrated a method of separation of actinium from francium and of francium from other actinides by elution with 10% ammonium carbonate solution. Rao and Gupta [476] investigated the radiocolloidal nature of actinium and suggested a method based on the fact that the colloid, if adsorbed on chromatographic paper, does not move with the solvent, whereas the ionic form does.

Radium

The most interesting problem in the chromatography of radium is its separation from barium. Although the methods of adsorption chromatography are the most efficient [409, 410] the problem can also be partly solved by paper chromatography if solvents containing mainly propanol are employed. Lithium nitrate added to solutions in nitric acid produces maximum separation, but the separation is limited to trace amounts [183]. For the separation of radium from its natural radioactive decay products, and for the separation of radium D from radium E, a mixture of acetone, $1M$ hydrochloric acid and $1M$ nitric acid (8 : 1 : 1) may be used [411]. Radium E can be purified chromatographically after its separation by extraction, to $85-95\%$ purity [412].

Protactinium

For the chromatographic separation of protactinium, Lederer used two solvents: with a mixture of acetone, hydrochloric acid and water, he separated protactinium from uranium, thorium, zirconium, titanium, vanadium, nickel, cobalt, beryllium, strontium, barium and aluminium [413], and with 1-butanol, hydrochloric acid and hydrofluoric acid mixture he isolated it from a mixture with niobium and tantalum [414]. Vernois [126] extended the application of the second solvent system for the separation of a number of other mixtures. Jakovac and Lederer [415, 416] used chromatography to study the properties of protactinium(V) in alkaline solutions [415, 416].

Transuranium Elements

Paper chromatography is also suitable for some separations in the transuranium series. Clanet [320, 417] investigated the possibility of separating plutonium(III), (IV) and (V) from americium(III) and uranium(IV) and (VI), using a mixture of 1-butanol and hydrochloric or nitric acid. Comparison of R_F values with those or uranium, thorium and some rare earths showed that this method is also suitable for separations of various combinations of the elements. Keller [418] investigated solvents containing methanol, ethanol, propanol or butanol, in combination with hydrochloric, nitric, or perchloric acid, and showed that it was possible to separate certain mixtures of actinium, thorium, protactinium, uranium, neptunium, plutonium and americium. Cowan and Foreman [419] and also Fink and Fink [420] paid special attention to plutonium, especially to the separation of oxidation states. Eschrich [421] separated uranium, plutonium, neptunium, and americium on silica gel, using reversed-phase chromatography.

The direct isolation of plutonium and subsequent liquid-scintillation counting of α-particles from ^{239}Pu was investigated by Crisco and Lada [655]. Various adsorption techniques, specificity and cross-contamination were studied in the presence of the fission products ^{89}Sr, ^{95}Zr, ^{106}Ru, ^{137}Cs and ^{144}Ce, and it was concluded that the strongly basic anion-exchange resin paper type SB-2 gives the best results for plutonium separation.

Phosphorus

Differentiating analytically between various phosphorus compounds is often difficult and usually requires a very good preliminary separation of the mixture to be analysed. Chromatographic methods are very suitable for this purpose, especially paper chromatography. They are used mainly in the analysis of mixtures of natural or artificial phosphates containing mainly dihydrogen orthophosphate and metaphosphate, polyphosphates, polymetaphosphates, diphosphates, phosphites, hypophosphites, etc. Greatest attention was devoted to the separation of these mixtures by paper chromatography by Ebel [428, 438, 441 — 450]. He proposed two types of solvent systems, acid and alkaline. The first are usually based on lower alcohols such as propanol, butanol or a mixture thereof, with the addition of trichloroacetic acid or picric acid; alkaline systems contain ammonia instead of acids. These solvents usually produce sufficiently sharp separation of single polyphosphates, and also of dihydrogen orthophosphate from diphosphate and from polymetaphosphates. Graham's salt as a rule remains on the start in all these solvents. Rössel and Kiesslich [556] have shown, however, that when methanol is used in the solvent system for the separation of condensed phosphates, triphosphates undergo hydrolysis. Good results have also been achieved with solvents containing dioxan. Kolloff [437] separated dihydrogen orthophosphate, diphosphate, tripolyphosphate, tetrapolyphosphate, trimetaphosphate, and tetrametaphosphate (in that order), using a dioxan — water mixture containing trichloroacetic acid and ammonia. Gauthier [438] used a solvent containing dioxan and boric acid, for a sharp separation of dihydrogen phosphate (R_F 0.4) from diphosphate (R_F 0.06). A mixture of equal volumes of dioxan, methanol and $1M$ ammonia has been used for the separation of orthophosphate, metaphosphate, pyrophosphate, phosphite and hypophosphite [436], but pyro and metaphosphate could not be separated. Ionova and Postnikov obtained the best reproducibility, developing the linear phosphates with 2-propanol, water, trichloroacetic acid and 25% aqueous ammonia solution (75 ml: 25 ml : 5 g : 0.3 ml) and separating the cyclic phosphates by development with 2-propanol, isobutyl alcohol, water and 25% aqueous ammonia solution (40 : 20 : 39 : 1) [656]. The separation of phosphate ions present in mixtures of inorganic cations was investigated by Lederer [440]. Using 1-butanol saturated with $1M$ hydrochloric acid, he was able to separate dihydrogen orthophosphate from various cations, in gram as well as microgram quantities. Phosphomolybdic acid and silico-molybdic acid have been separated with 2-propanol—water—acetic acid (25 : 5 : 2) and the spots identified by spraying with selective reagents [699].

The separated phosphates can be determined quantitatively in several ways. Comparison with a standard chromatogram or planimetry of the spots is difficult, in view of the nature of the spots. Therefore, colorimetric methods are most suitable. After elution of the spot from the paper the corresponding compound is hydrolysed to dihydrogen orthophosphate and the latter determined as phosphomolybdenum blue. For trace amounts of separated compounds, labelled phosphates (^{32}P) may be used.

In Table 6.5 a review of selected separations of phosphorus compounds is given.

Sulphur

Polythionates are the inorganic sulphur compounds most often separated chromatographically. Pollard [452, 453] proposed a mixture of 1-butanol and 2-propanol with acetone and potassium acetate, as these mixtures separate sulphate, thiosulphate, dithionate and higher thionates up to hexathionate. For the separation of polythionates Scoffone [457] proposed a mixture of 1-butanol, acetic acid and ethyl acetate. Bighi and co-workers [458] separated polythionates up to $[S_{12}O_6]^{2-}$ in a mixture of 1-butanol, acetic acid and water. Coch-Frugoni [459] investigated the possibility of chromatographic separation of sulphuric acid, nitric and phosphoric acids and found that sharp separation can be achieved in 1-butanol saturated with $1M$ hydrochloric acid in the following increasing order of R_F values: H_2SO_4, H_3PO_4, HNO_3. Servigne and Duval [451] separated various compounds of sulphur, using a 3 : 1 mixture of 1-butanol and water, or 2-propanol and water.

For the identification of thiosulphate in the presence of chloride, sulphate, carbonate, nitrate, sulphite and chromate, separation on paper impregnated with silver chloride, with water as solvent, can be used [557].

Pollard and co-workers made use of paper chromatography for the study of certain processes in inorganic chemistry. They studied mainly the conditions of formation and the reactions of polythionates [454] and quantitatively analysed mixtures containing thiosulphate and polythionates [455]. They also followed the reactions of polythionates in acid medium [452] and studied chromatographically the oxidation of sulphurous acid with ferric ions [459].

Table 6.5. Review of Some Separations of Phosphorus Compounds

Compound	Solvent							
	1	2	3	4	5	6	7	8
PO_4^{3-}	0.65	0.73	0.80	0.70	0.79	0.21	0.60	0.34
$P_2O_7^{4-}$	0.44	0.53	0.59	0.50	0.68	0.15	—	0.20
$P_3O_{10}^{5-}$	0.29	0.39	0.45	0.38	0.58	—	—	—
$P_4O_{13}^{6-}$	0.17	0.29	0.35	0.25	0.47	—	—	—
$P_5O_{16}^{7-}$	0.11	0.22	0.26	0.18	0.36	—	—	—
$P_6O_{19}^{8-}$	0.07	0.16	0.18	0.13	0.25	—	—	—
$P_7O_{22}^{9-}$	0.04	0.11	0.11	0.09	0.15	—	—	—
$P_8O_{25}^{10-}$	—	0.08	—	—	—	—	—	—
$P_3O_9^{3-}$	0.20	0.32	—	0.21	0.39	—	—	—
$P_4O_{12}^{4-}$	0.08	0.18	—	0.13	0.22	—	—	—
PO_3^{-}	—	—	—	—	—	0.0	—	—
PO_3^{2-}	—	—	—	—	—	—	0.85	0.56
PO_3^{3-}	—	—	—	—	—	—	0.71	—
$P_2O_6^{4-}$	—	—	—	—	—	—	0.44	—
PO_3S^{3-}	—	—	—	—	—	—	—	—
HPO_2S^{2-}	—	—	—	—	—	—	—	—
H_2POS^{-}	—	—	—	—	—	—	—	—
H_2PS_2	—	—	—	—	—	—	—	—
References	[429]	[429]	[430]	[431]	[431]	[432]	[433]	[105]

Solvent compositions:

1 — iPrOH + H_2O + CCl_3COOH + 20% NH_3 (750 : 250 : 50 : 3 [428]).
2 — iPrOH + H_2O + CCl_3COOH + 25% NH_3 (700 : 100 : 200 : 3).
3 — iPrOH + CCl_3COOH + NH_3 + H_2O + ethyleneglycol monomethyl ether (800 : 50 : 3 : 400 : 400).
4 — 26.25 ml of 2-propanol, 13.75 ml of water, 15.6 ml of trichloroacetic acid and 5.5 ml of concentrated ammonia solution, made up to 100 ml with water, 3 ml of acetic acid (20 ml of glacial acetic acid + 80 ml of water) and 30 ml of dioxan.
5 — 60 ml of methanol, 10.3 ml of trichloroacetic acid solution (100 g of trichloroacetic acid dissolved in 500 ml of water plus 22.7 ml of concentrated ammonia solution) and 5 ml of acetic acid solution (20 ml of glacial acetic acid + 80 ml of water)
6 — BuOH + AcOH + H_2O (4 : 1 : 5).
7 — iPrOH + iBuOH + H_2O + 25% NH_3 (60 : 30 : 95 : 15).
8 — iBuOH + iPrOH + NH_3 + H_2O (20 : 40 : 1 : 39).

In Table 6.6 (see p. 120) a review of some separations of sulphur compounds is given.

Halogens

When choosing suitable solvent systems for the separation of halogen anions attention must be paid to the oxidation states. Probably the most suitable solvent mixture for the separation of chloride, bromide, iodide and

Table 6.5 continued

Compound	Solvent								
	9	10	11	12	13	14	15	16	17
PO_4^{3-}	0.17	0.91	0.76	0.72	0.68	0.43	0.45	0.4	0.60
$P_2O_7^{4-}$	—	0.82	0.56	0.55	0.49	0.33	0.36	0.06	—
$P_3O_{10}^{5-}$	—	—	—	—	—	—	—	—	—
$P_4O_{13}^{6-}$	—	—	—	—	—	—	—	—	—
$P_5O_{16}^{7-}$	—	—	—	—	—	—	—	—	—
$P_6O_{19}^{8-}$	—	—	—	—	—	—	—	—	—
$P_7O_{22}^{9-}$	—	—	—	—	—	—	—	—	—
$P_8O_{25}^{10-}$	—	—	—	—	—	—	—	—	—
$P_3O_9^{3-}$	—	0.55	0.34	0.44	0.34	0.54	0.74	—	—
$P_4O_{12}^{4-}$	—	0.44	0.22	0.38	0.25	0.48	0.61	—	—
PO_3^-	0.04	—	—	—	—	—	—	—	—
PO_3^{2-}	0.69	—	—	—	—	—	—	—	—
PO_3^{3-}	0.36	—	—	—	—	—	—	—	—
$P_2O_6^{4-}$	—	—	—	—	—	—	—	—	—
PO_3S^{3-}	—	—	—	—	—	—	—	—	0.53
HPO_2S^{2-}	—	—	—	—	—	—	—	—	0.75
H_2POS^-	—	—	—	—	—	—	—	—	0.64
H_2PS_2	—	—	—	—	—	—	—	—	0.33
References	[436]	[437]	[439]	[439]	[439]	[439]	[439]	[438]	[456]

Solvent compositions:

9 — $MeOH + dioxan + 1\ NNH_3$ (1 : 1 : 1).
10 — $dioxan + H_2O + CCl_3COOH + NH_3$ (240 : 140 : 20 : 1).
11 — $iPrOH + H_2O + CCl_3COOH$ (70 : 30 : 5).
12 — $tBuOH + iPrOH + H_2O + CCl_3COOH$ (40 : 30 : 30 : 5).
13 — $tBuOH + H_2O + picric\ acid$ (80 : 20 : 4).
14 — $iPrOH + iBuOH + NH_3 + H_2O$ (40 : 20 : 1 : 39).
15 — $PrOH + EtOH + H_2O + NH_3$ (30 : 30 : 39 : 1).
16 — $dioxan + iPrOH + H_2O + H_3BO_3$ (300 : 100 : 300 : 1).
17 — $iPrOH + H_2O + CCl_3COOH + NH_3$ (800 : 200 : 50 : 3).

fluoride is acetone–pyridine–water [462] with which all four anions can be very sharply separated. 1-Butanol saturated with $1.5M$ ammonia [464] can also be used, as it separates the four halides from each other and also from thiocyanate. When circular chromatography is used, chloride, bromide, iodide and thiocyanate can be sharply separated with a mixture of methanol, acetone and aqueous ammonia (3 : 14 : 4) [558]. Chloride, bromide, sulphate and phosphate can be separated by various inorganic solvents on papers impregnated with ferric hydroxide [463]. For the separation of chlorate, bromate and iodate and for their partial separation from perchlorate, chlorite and hypochlorite, 75% aqueous propanol is most suitable [460, 461]. A mixture of 1-butanol, pyridine and ammonia will separate chloride, bro-

Table 6.6. Review of Some Separations of Sulphur Compounds

Compound	tBuOH + Me_2CO + H_2O + KOAc (30:130:50:1)	iPrOH + Me_2CO + H_2O + KOAc (50:20:30:2)	BuOH + AcOH + EtOAc + H_2O (10:2:1:7)	BuOH + AcOH + H_2O (4:1:5)	BuOH + H_2O (3:1)	iPrOH + H_2O (3:1)
SO_4^{2-}	0.02	—	—	—	0.0	0.43
S^{2-}	0.04	—	—	—	0.0	0.32
$S_2O_3^{2-}$	0.11	0.10	—	—	0.0	0.35
$S_2O_6^{2-}$	0.38	—	—	—	0.0	0.38
$S_3O_6^{2-}$	0.67	0.32	0.20	0.06	—	—
SO_3^{2-}	0.77	—	—	—	—	—
$S_2O_8^{2-}$	0.89	—	—	—	0.36	0.50
$S_4O_6^{2-}$	1.00	0.45	0.30	0.12	0.02	0.39
$S_5O_6^{2-}$	1.25	0.74	—	0.21	—	—
$S_6O_6^{2-}$	1.43	0.82	0.48	0.28	—	—
$S_7O_6^{2-}$	—	—	—	0.32	—	—
$S_8O_6^{2-}$	—	—	—	0.37	—	—
$S_9O_6^{2-}$	—	—	—	0.41	—	—
$S_{10}O_6^{2-}$	—	—	—	0.53	—	—
$S_{11}O_6^{2-}$	—	—	—	0.95	—	—
SO_3^{2-}	—	—	—	—	0.03	0.11
$S_2O_5^{2-}$	—	—	—	—	0.0	0.34
HSO_4^-	—	—	—	—	0.0	0.33
$S_2O_7^{2-}$	—	—	—	—	0.57	0.51
Ref.	[452]	[453]	[457]	[458]	[451]	[451]

Note: R_F values relative to the R_F value of $S_4O_6^{2-}$ taken as 1.00.

mide and iodide, as well as chlorate, bromate and iodate [212]. The latter group can also be separated well with a mixture of 1-butanol, acetone and ammonia (1 : 3 : 1) or 1-butanol, dimethylformamide and ammonia (3 : 1 : 1) [559]. Ando and Ishii [466] separated chloride, bromide and iodide very sharply with a mixture of 1-butanol and picoline at a temperature of 5 to 10 °C. A mixture of chloride, chlorite, chlorate and perchlorate can be separated with a mixture of 2-propanol, water, pyridine and concentrated ammonia solution (15 : 2 : 2 : 2) [561]. Thielemann [700] has separated hypochlorite, chlorite, chlorate and perchlorate by paper and thin-layer chromatography.

Impregnated papers (usually with silver salts) have been used for the separation of chloride, bromide and iodide. Papers impregnated with silver nitrate [467], chromate [560, 468], or arsenate [469] give good separation if aqueous ethanol is used as solvent. The separation of a mixture of halides is based on the selective precipitation of silver halides, depending on the solubility product. The reaction area on the chromatogram is usually proportional to the amount of the anion. Iodine or its compounds can be quantitatively determined on paper chromatograms by short activation in a beam of thermal neutrons and subsequent estimation of the induced radioactivity by means of a Geiger-Müller counter [470]: as little as 0.02 µg of iodine can be determined. For quantitative determination of microgram amounts of fluoride on filter paper Kielczewski and co-workers [657] recommended ascending chromatography on paper impregnated with the Al-aluminon complex, with propanol, water and ammonium acetate buffer as developing system. The areas of the white spots are proportional to amount of fluoride in the range $1-7$ µg. There is no interference from other halides but a fresh calibration graph is required with each batch of paper.

A review of some separations of halogen compounds is given in Table 6.7.

Table 6.7. Review of Some Separations of Halogen Compounds

Ion	Solvent							
	1	2	3	4	5	6	7	8
F^-	—	0.0	—	—	—	0.0	—	—
Cl^-	0.46	0.20	0.8	0.90	0.10	0.23	0.24	—
Br^-	—	0.36	—	—	0.17	0.47	0.36	—
I^-	—	0.69	0.52	0.48	0.30	0.72	0.47	1.0
ClO_3^-	0.68	—	—	—	—	—	0.42	—
BrO_3^-	0.48	—	—	—	—	—	0.25	—
IO_3^-	0.20	—	—	—	—	—	0.09	0.72
ClO_4^-	0.65	—	—	—	—	—	—	—
IO_4^-	—	—	—	—	—	—	—	0.0
ClO_2^-	0.50	—	—	—	—	—	—	—
ClO^-	0.55	—	—	—	—	—	—	—
SO_4^{2-}	—	—	0.60	0.8	—	—	0.07	—
PO_4^{3-}	—	—	0.0	0.2	—	—	0.04	—
SCN^-	—	—	—	—	0.45	—	0.56	—
Ref.	[460, 461]	[462]	[463]	[463]	[464]	[465]	[212]	[471]

Solvent compositions:

1 — iPrOH + H_2O (3 : 1).
2 — Me_2CO + PY + H_2O (4 : 2 : 1).
3 — $0.05M$ $NaHCO_3$; paper impregnated with $Fe(OH)_3$.
4 — $0.1M$ NaOH; paper impregnated with $Fe(OH)_3$.
5 — BuOH saturated with $1.5M$ NH_3.
6 — PY + H_2O (9 : 1).
7 — BuOH + PY + $1.5M$ NH_3 (2 : 1 : 2).
8 — BuOH + EtOH + $5M$ NH_3 (1 : 1 : 1).

Other Anions

For the separation of a mixture of inorganic anions several solvent systems have been proposed. Nakano [482] recommends a mixture of ammonia, acetone and 1-butanol in which the R_F values increase in the following order: MnO_4^-, CrO_4^{2-}, SO_4^{2-}, $S_2O_3^{2-}$, CO_3^{2-}, CN^-, Cl^-, Br^-, I^-, CSN^-. A mixture of butanol, pyridine and $1.5M$ ammonia [212] separates chloride, bromide and iodide; chlorate, bromate and iodate; nitrite and nitrate; arsenite and arsenate. The R_F values increase as follows: CrO_4^{2-}, (PO_4^{3-}, AsO_4^{3-}, CO_3^{2-}, SO_4^{2-}, IO_3^-), AsO_3^{3-}, Cl^-, (NO_2^-, BrO_3^-), Br^-, (NO_3^-, ClO_3^-), I^-, SCN^-. A mixture of ethanol, pyridine and ammonia [472] also gives separation of nitrite and nitrate, sulphate and sulphite, and sulphates from chloride and nitrate. A mixture of 2-propanol and $1.5M$ ammonia [479] separates ferrocyanide from ferricyanide, arsenite from arsenate, chlorate, bromate and iodate, perchlorate and periodate, and also gives certain separations between these groups of anions. Inorganic solvents are also important in the paper chromatography of inorganic anions. Thus, an aqueous solution of ammonium sulphate separates chloride, bromide and iodide, as well as halide from thiocyanate [481]. An aqueous solution of sodium acetate and acetic acid [478] separates ferrocyanide and ferricyanide from halides and chromate. Pure water was used for the determination of cyanides on papers impregnated with silver chromate, chloride or bromide [562].

Chromatographic methods are also suitable for the separation of compounds of boron, and also, to a certain extent, those of silicon. Borates can be separated from tetraborates with 80% aqueous ethanol [477]. The use of chromatographic methods for the separation and for the determination of silicates is, however, limited only to special cases, as for example the separation of silicates from borate [332] or phosphate [333]. Silicates usually remain on the start in acetone – hydrochloric acid solvents, in which the majority of other anions have R_F about 0.9 and borate about 0.6. With sodium borate and hydrochloric acid as solvent, phosphorus can also be separated, because it moves with the solvent front. Baumann [334] used a mixture of 2-propanol, water and acetic acid for the chromatographic study and separation of silicic acids. Monosilicic acid has R_F 0.75, while the polymolecular acids formed on condensation have R_F values up to 1.0. Malý [563] recommends chromatographic methods for the determination of silica in ash from biological material. The ash is fused with a sodium and potassium carbonate mixture (for total silica) or potassium bicarbonate and chloride mixture (for free silica), and the cooled melt is leached with water and the extract chromatographed with a mixture of ethanol and saturated strontium hydroxide solution (25 : 23). The faintly fluorescent spots of strontium metasilicate at R_F 0.60 are then estimated semiquantitatively by planimetry. Silicomolybdic acid can be separated from phosphomolybdic acid by chromatography with 2-propanol – water – acetic acid (20 : 5 : 2) [699]. Certain separations can be carried out on impregnated papers. Kielczewski and co-workers recommended, for example, papers impregnated with silver nitrate [564] or cupric sulphide [565] for the separation of cyanide from other ions. With the first method, cyanide and iodide can be determined simultaneously, and with the second, cyanide may be determined in the

Table 6.8. Review of Anion Separations in Various Solvent Systems

Anion	Solvent										
	1	2	3	4	5	6	7	8	9	10	11
NO_2^-	0.25	0.43	—	—	—	0.43	0.85	0.49	0.41	0.63	0.74
NO_3^-	0.40	0.56	—	—	—	0.45	0.83	—	—	—	0.72
AsO_3^{3-}	0.19	—	—	—	0.86	0.11	0.70	0.84	0.79	0.79	—
AsO_4^{3-}	0.05	0.03	0.26	—	0.98	0.01	0.83	0.89 to 1.0	0.48	0.88	—
AsO_2^-	—	0.12	—	—	—	—	—	—	—	—	—
PO_4^{3-}	0.04	0.03	0.37	—	1.0	0.02	0.80	0.90 to 1.0	0.42	0.92	—
CrO_4^{2-}	0.00	0.03	—	—	0.64	0.03	0.83	0.18 to 0.57	0.0	0.0 to 0.17	—
$Cr_2O_7^{2-}$	—	0.03	—	—	—	—	—	—	—	—	—
CSN^-	0.56	0.66	—	—	0.77	0.59	0.81	0.07	0.16	0.10	0.39
CN^-	—	0.06	—	—	—	—	—	—	—	—	—
SO_4^{2-}	0.07	0.07	—	—	—	0.04	0.84	—	—	—	—
$S_2O_3^{2-}$	—	0.09	—	—	—	—	—	0.75	0.10	0.83	—
SO_3^{2-}	—	0.20	—	—	—	0.05	0.77	—	—	—	—
S^{2-}	—	0.03	0.16	—	—	—	—	—	—	—	—
F^-	—	0.01	—	—	—	0.24	0.78	—	—	—	—
Cl^-	0.24	0.45	—	—	0.87	0.32	0.88	0.57	0.52	0.66	0.74
Br^-	0.36	0.50	—	—	0.86	0.36	0.85	0.35	0.43	0.43	0.64
I^-	0.47	0.61	0.92	—	0.83	0.48	0.81	0.10	0.24	0.12	0.46
ClO_3^-	0.42	0.60	—	—	—	0.47	0.85	0.48	0.45	0.49	0.80
BrO_3^-	0.25	0.41	—	—	—	0.27	0.81	0.74	0.49	0.75	0.81
IO_3^-	0.09	0.10	—	—	—	0.04	0.76	0.95	0.33	0.88	0.81
ClO_4^-	—	—	—	—	—	0.60	0.89	—	—	—	—
IO_4^-	—	—	—	—	—	0.04	0.77	0.06	0.0	0.06	—
CO_3^{2-}	0.06	0.10	—	—	—	—	—	—	—	—	—
$Fe(CN)_6^{4-}$	—	0.01	0.12	—	1.0	0.02	0.94	0.50	0.0	0.43	—
$Fe(CN)_6^{3-}$	—	0.08	0.15	—	1.0	0.15	0.93	0.0	0.0	0.0	—
BO_3^{3-}	—	—	—	0.65	0.87	0.13	0.56	—	—	—	—
$B_4O_7^{2-}$	—	0.09	—	0.16	—	—	—	—	—	—	—
Ref.	[212]	[472]	[473]	[477]	[478]	[479]	[479]	[480]	[480]	[480]	[481]

Solvent compositions:
1 — BuOH + PY + 1.5M NH$_3$ (2 : 1 : 2).
2 — EtOH + PY + NH$_3$ + H$_2$O (15 : 5 : 1 : 4).
3 — BuOH + EtOH + H$_2$O (2 : 2 : 1).
4 — EtOH + H$_2$O (4 : 1).
5 — NaOAc (10 g) + AcOH (10 ml) + H$_2$O (to make 100 ml).
6 — iPrOH + 1.5M NH$_3$ (4 : 1).
7 — iPrOH + 1.5M NH$_3$ (1 : 4).
8 — 1M KNO$_3$; Whatman weakly alkaline exchange paper.
9 — 1M KNO$_3$; Amberlite WB-2 paper.
10 — 1M KNO$_3$; Amberlite SB-2 paper.
11 — 8N (NH$_4$)$_2$ SO$_4$.

Table 6.9. Review of Some Chromatographic Separations
on Cellulose Columns

Ion	Solvent	Remark	Ref.
Al	ethyl methyl ketone + HCl	separation from other ions in steel	[483]
Au	ethyl acetate + + HNO_3	separation from a series of ions, including Pt-metals	[375]
Ba, Ra	MeOH + HCl	mutual separation	[485]
Ca, Sr	MeOH + Et_2O + + HCl	mutual separation	[484]
Ba, Sr	MeOH + HCl MeOH + Et_2O + + HCl	mutual separation	[485]
Mg, Ca, Sr, Ba, Ra	MeOH + HCl	microcrystalline cellulose	[709]
Ge	pyridine	separation from Sb	
Pt metals	MeiBuCO + HCl MeiBuCO + HCl + + $NaClO_3$	separations: Ir—Rh, Rh from the others	[487]
Nb, Ta	ethyl methyl ketone ethyl methyl ketone + + HF	separation from other ions mutual separation	[132] [134 to 136]
Mo	acetylacetone	separation from other ions in steel	[488]
Sn, Cu	*n*-BuOH + HCl	mutual separation	[490 to 493]
Th, U	Et_2O + HNO_3	separation from other ions and mutual separation	
U	$0.2M$ Na_2CO_3 $0.05M$ NH_4HCO_3	in the form of a carbonate complex from a series of other ions	[321]
Zn	*n*-BuOH + HCl	separation from copper alloys	[495]
Cd	Et_2O + 5% HNO_3	quantitative determination by spectrography	[568]
Rare earths Au	Et_2O + HNO_3	separation from Ag, Fe, Zn, Cd, Mn, Co	[569]
Sb	HCl in methanol	separation from Mn, Fe, Au, U, Hg, As and others	[570]
Sb	Aqueous HNO_3 or methanolic HCl	Separation from Mn, Fe, Au, U, Hg, As	[689]
Trace metals	Et_2O + HNO_3	trace In, Zr, Co, Cu, Cr, Mn, Ba, Fe, and Ag in $UO_2(NO_3)_2$	[571]

Table 6.10. Review of Chromatographic Separations on Silica Gel Columns

Ion	Solvent	Remarks	Ref.
Fe	conc. H_2SO_4	determination of traces of Fe in conc. H_2SO_4	[48]
Cd, Ag	from KCN solution	separation from other ions	[496]
U, Np, Pu, Am	Bu_3PO_4 + HNO_3	mutual separation	[421]
Zr, Hf	methanol	elimination of Hf from Zr	[45]
U	aqueous solution of pH 4	retention in the presence of EDTA	[54, 55]
Be	aqueous solution of pH 5.5—7.5	retention in the presence of EDTA	[56]
Sn	aqueous solution of pH 5—10	separation from Sb in the presence of citric acid	[57, 58]
Fe	aqueous solution of pH 5.5—6	column with 1,10-phenanthroline	[49]
Hg	chloroform	column with dithizone	[51]
Cu, Ni, Co	chloroform + ethanol	oxinates	[52]
Cu	aqueous HNO_3 (1 : 2)	column with CdS	[497]
V	aqueous solution of pH 1.4—6.7	column with 8-hydroxyquinoline	[53]
Rare earths	tributyl phosphate saturated with $13M$ HNO_3	mutual separation	[47]
Sn, Te, Sb	tributyl phosphate	$Sn^{2+}-Sn^{4+}$; $Sn^{2+}-Te^{4+}$ and Sb^5	[422]
In	aqueous $HClO_4$	separation from cadmium; column with dithizone	[458]
U	methyl isobutyl ketone	separation in $6M$ HNO_3 from many other ions	[572]
Nb	stationary phase: tributyl phosphate, elution with conc. HCl	separation from a large excess of molybdenum	[573]
Ho, Dy	stationary phase: di-(2-ethylhexyl)phosphate elution with $0.1—4M$ HCl	traces of Ho in Dy and traces of Lu in Yb	[575]
Y	stationary phase: di-(2-ethylhexyl)phosphate, elution with $1.2M$ HCl		[574]
Np(IV), Np(V) Np(VI)	$0.5—2M$ HNO_3	silica gel impregnated with tributyl phosphate	[576]
Alkali metals	$0.01M$ HNO_3 $0.04M$ LiCl $0.01M$ HCl	silica gel impregnated with 2,6,8-trimethylnonyl orthophosphate	[582]
Rare earths	HNO_3	silica gel impregnated with di-(2-ethylhexyl) orthophosphate	[583]

Table 6.11. Review of Chromatographic Separations on Columns
of Various Materials

Ion	Solvent	Remarks	Ref.
rare earths	aqueous HCl, $HClO_4$ others	column with di(2-ethylhexyl)-orthophosphoric acid; mutual separation	[498–503]
rare earths	$4M$ HNO_3; $8M$ HCl; or $11.5-13M$ HNO_3	column of tributyl phosphate	[504]
rare earths	HCl and HNO_3, of various concentrations	column of silica gel impregnated with 2-ethylhexylphosphoric acid	[508]
Bi, Hg, Ag	HCl of various concentrations	column of "Fluoroplast" and tributyl phosphate	[426]
Mn, Co, Ni, Cu—Fe—Zn, Ca, Sr, Y—Fe—Cd	HCl of various concentrations	column of Teflon with tri-n-octyl-amine	[505]
Th, Pa, —U, Ni, Co, Fe	$4M$ HCl	column of tri-n-octylamine	[506]
Th, Zr, U	$10M$ HCl; $10M$ NH_4NO_3	column of tri-n-octylamine	[506]
Zr, La	$10M$ NH_4NO_3	column of tri-n-octylamine	[506]
Hf, Zr	$8M$ HCl + 5% HNO_3	column of tri-n-octylamine	[506]
La, U, Th	$10M$ NH_4NO_3	column of tri-n-octylamine	[506]
Fe, Co, Ni	$7M$ HCl; $6M$ HCl; $0.5M$ H_2SO_4	cellulose impregnated with tri-n-octylphosphine	[507]
U, Th, rare earths	$8M$ HCl; $1M$ HCl; $4M$ H_3PO_4	cellulose impregnated with tri-n-octylphosphine	[507]
La, Th, Zr	$8M$ HCl; $1M$ HCl; $4M$ H_3PO_4	cellulose impregnated with tri-n-octylphosphine	[507]
Al, Cu, Fe, U	$6M$ HCl; $2M$ HCl; $0.5M$ H_2SO_4; $4M$ H_3PO_4	cellulose impregnated with tri-n-octylphosphine	[507]
Pb, Bi, Fe	$6M$ HCl; $10M$ HCl; $0.5M$ H_2SO_4	cellulose impregnated with tri-n-octylphosphine	[507]
U	$5.5M$ HNO_3	polychlorotrifluoroethylene ("Kel-F") impregnated with tri-n-butyl phosphate	[510]
	HNO_3 of various concentrations	styrene-divinylbenzene, tributyl phosphate	[511]
Cs	$0.01M$ EDTA; $1M$ HCl acetone	"Kel-F", tetraphenylborate	[512]
Sn	$8M$ HCl	Teflon impregnated with methyl isobutyl ketone. Separation from Bi, Cd, Cu, Pb, Hg, Zn	[577]
Zr, Hf	$SCN^-$$(NH_4)_2SO_4$	Teflon with stationary methyl isobutyl ketone	[578, 579]

Table 6.11 continued

Ion	Solvent	Remarks	Ref.
Cf, Cm, Bk	dilute HNO_3	Teflon with stationary di-(2-ethyl-hexyl)orthophosphoric acid	[580]
Alkali metals	$3M$ HNO_3	"Kel–F" impregnated with bis (dihexylphosphinyl)methane	[581]
Am, Y, rare earths	$0.6M$ NH_4SCN	"Plaskon" (polymer of trifluoro-chloroethylene) impregnated with Aliquat 336 (tricapryl-methylammonium chloride)	[584]
Sn	$1M$ HCl, $3M$ H_2SO_4, $8M$ HCl	Teflon-6 column. Methyl iso-butyl ketone as stationary phase Separation from Bi, Cd, Cu, Hg(II), Pb, Zn	[690]
Bi	$0.5M$ HNO_3	Polytetrafluorethylene column impregnated with di-isoamyl hydrogen phosphate, Separation from Ni, Co, Cu, Fe, Cr, Mn, Zn, Pb, As, Al, Sn, Mg etc.	[691]
Rare earths	$0.1M$ HCl	Bis(2-ethylhexyl)phosphate supported on water-repellent silica gel column	[692]
Am(V)	$0.01M$ HNO_3	Teflon-6 impregnated with bis-(2-ethylhexyl) hydrogen phos-phate in heptane solution. Separation from Cm, other actini-des, lanthanides and many other metals	[693]
Ac, La, Ba	$10M$ NH_4NO_3 in $0.1M$ NHO_3	Tributyl phosphate on Fluoro-plast-4 column	[694]
Nb, Ta, Mo, W	various aqueous mix-tures of HCl and HF	Teflon-6 column impregnated with iodine	[695]
Pu	HNO_3 at various conc.	Purified tributyl phosphate sup-ported on poly(chlorotrifluoro-ethylene). Pu retained on the column, most of the common inpurities remo-ved	[696]
U	90% $Al(NO_3)_3 . 9 H_2O$	Tributyl phosphate on silica gel column. Separation from pho-sphoric acid. U eluted with water	[697]
Pu(III), U(IV), U(VI)	HNO_3 + hydrazine	Tributyl phosphate on poly(chlor-otrifluoroethylene)	[701]
Cu		α-Hydroxyoximes on solid sup-port	[702]

Table 6.11 continued

Ion	Solvent	Remarks	Ref.
Zn, Cd, Pb, Cu		Chitosan	[703]
Hg, Co, Au, Sb, Ag, Cr, Fe, U, Cs, Zn, Ir, Pd, Cu, Mo, SO_3^{2-}		Chitin or chitosan	[704]
U, Pd, Au,	$1M$ LiClO$_4$/ 0.025M HClO$_4$	Dioctyl sulphoxide on polystyrene support	[705]
General examination of behaviour of metal ions		Various sulphoxides on polystyrene support	[706]
Ga, In, Tl	HBr, HNO$_3$, isopropyl ether, MeiBuCO	Polystyrene impregnated with isopropyl ether of MeiBuCO. Gallium only partially sorbed on column (20—30%)	[707]
Bk	$8M$ HNO$_3$/0.2M NaBrO$_3$	Siliconed diatomaceous earth treated with di(2-ethylhexyl) hydrogen phosphate. Separation from lanthanides and actinides	[708]
U(VI), U(IV)	HNO$_3$ hydrazine	Columns treated with di(2-ethylhexyl)hydrogen phosphate. Separation from fission products, Pu, heavy metals	[711]
Cu		Colums treated with substituted oximes. Separation from Fe, Co, Ni, Mn, Cr, Mo, W, V	[712]
Np, U, Pu		Kel-F impregnated with trilanoylammonium nitrate	[716, 717]
Th, U, Y, I		Microporous polyethylene treated with tri-n-octyl phosphine oxide (for U, Th), di-(2-ethylhexyl)-orthophosphoric acid (for y) or benzene (for I)	[747]

presence of an excess of sulphate, thiocyanate, cyanate, chloride, bromide and iodide. The area of the spots is proportional to the amount of cyanide present. Microgram amounts of iodide and bromide can be determined in the usual manner on papers impregnated with silver nitrate [566], and nitrite and nitrate may be determined indirectly on paper impregnated with iodine—starch complex [567]. Thielemann [701] has used methanol—pyridine—dioxan (7 : 2 : 1) to separate cyanide, thiocyanate, cyanate and carbonate. The R_F values of anions in some solvent systems are given in Table 6.8.

6.2.2. Chromatography on Cellulose Columns

The principles of partition chromatography on columns of cellulose and silica gel have already been described in Section 2.2.2. Here a few separations best performed by this technique will be mentioned. In Table 6.9 examples are given of such separations, developed, as a rule, for the analysis of definite materials. Particular applications will be described in more detail in Chap. 7.

6.2.3. Chromatography on Silica Gel Columns

Silica gel can be used alone or as carrier for an organic stationary phase (for use with a suitable organic solvent). Silica gel is often made hydrophobic by siliconing with silane before coating with the organic stationary phase. In Table 6.10 some examples of separations are given.

6.2.4. Chromatography on Columns of Various Materials

Sometimes it is difficult to classify a method as partition chromatography or as some other chromatographic method, such as ion-exchange. Cellulose, silica gel, or even polyvinyl chloride, Teflon, etc. are often impregnated with solvent to form the stationary phase (reversed-phase chromatography) and thus act as liquid ion-exchangers. As in Section 6.2.1, where these methods were counted among paper chromatographic methods, some of them are quoted here as examples of special, often very efficient, methods of column chromatography. Examples of separations are reviewed in Table 6.11.

6.2.5. Thin-Layer Chromatography

For separation on thin layers, silica gel and powdered cellulose are mainly used. In the past few years many applications of this method have been developed, covering a broad field of inorganic analysis. The literature gives examples of separations of simple and complex ions of metals and non metals. Various separations of anions have been made, and reversed-phase-chromatography has also been successfully applied.

Seiler and co-workers were among the first to develop thin-layer chromatography for the analysis of inorganic mixtures [590, 599]. In particular they investigated the separation of uranium from gallium, separations within the group of alkali metals, and separation of halides, phosphorus-containing acids, and polythionates. They also studied the principles of separation on thin layers and concluded that the main factors influencing the separation are the ion-exchange property of the adsorbent and the co-ordinative properties of the solvents. They also studied the *cis-trans* isomerism of cobalt salts and the problem of determination of ions after their separation on thin layers. Hashmi and Chughtai [728] have published a method of semiquantitative estimation of 30 cations out of 46 in the scheme, based on circular TLC after subdivision into groups by solvent extraction. Some of the more important separations are presented in the following short review. The applications of TLC in inorganic analysis were reviewed in 1970 by Volynets and Ermakov [729].

Alkali metal ions can be separated in the form of their polyiodides. Layers of moist silica gel are used with iodine in nitrobenzene as the mobile phase; the most suitable method of development is in a 'sandwich' chamber. If the concentration of iodine in the solvent system is $0.1M$ the following sequence of R_F values is observed: $Cs^+ > Rb^+ > K^+ > NH^+ > Na^+ > Li^+$ [600]. The autoradiographic method can be used for their identification after the separation of ^{24}Na, ^{42}K, ^{86}Rb and ^{137}Cs on layers of cellulose powder containing 10% of zinc ferrocyanide, with $0.2M$ ammonium nitrate as solvent [658]. The alkali metals have been separated from bi- and tervalent metals by TLC [718]. Alkaline earth metals can be separated on a layer of powdered cellulose by a mixture of dioxan, hydrochloric acid and water, or methanol, hydrochloric acid and water. Single elements can be determined in amounts down to 0.1 µg by detection with 8-hydroxyquinoline [601]. The method has recently been modified [719]. On a layer of powdered cellulose, a group of these metals can be separated from beryllium, magnesium and barium for quantitative determination of these three elements. Photometric methods are the most suitable. For the determination of beryllium 8-hydroxyquinoline is used, for magnesium, Eriochrome Black T, and for barium, metalphthalein [602]. Separations of $^{90}Sr/^{90}Y$, $^{140}Ba/^{140}La$, $^{131}Ba/^{131}Cs$ and $^{137}Cs/^{137}Ba$ have been reported [718]. For the determination of nickel in the presence of other ions of the ammonium sulphide group, Gagliardi and Pokorný [603] proposed separation with acetone, hydrochloric acid and water (75 : 13 : 12), using a cellulose thin layer. Nickel moves with the solvent front and can be well separated from aluminium, titanium, vanadium, manganese, iron, cobalt, zinc, zirconium and uranium. After its extraction from the thin layer, nickel can be determined photometrically by adding 8-hydroxyquinoline and extracting the complex with chloroform. Copper, cobalt and nickel are separated from a large excess of up to 40 other metals by shaking the sample solution with a mixture of chloroform, pyridine and 50% aqueous ammonium thiocyanate solution (20 : 1 : 2). The chloroform layer is then removed and subjected to circular thin-layer chromatography on alumina with acetone–acetylacetone–$4M$ hydrochloric acid (91 : 4 : 5). The spots of the three metal ions can be evaluated quantitatively by spectrophotometric methods [659]. The metals in the ammonium sulphide analytical group can be separated on layers of cellulose by use of propanol – $6M$ hydrochloric acid (3 : 1) as solvent mixture. After extraction of the adsorbent with $0.1M$ hydrochloric acid, iron can be determined photometrically by reaction with 1% morpholinium morpholine–N–dithiocarbamate and extraction of the complex into chloroform [660]. Nickel, iron, copper and cobalt can be separated by combined ring-oven and TLC on alumina–coated plates, the pyridine–2-aldehyde–2-quinolylhydrazine complexes being extracted into chloroform and chromatographed with chloroform on the ring-oven [710]. The salicyl-aldoxime complexes of nickel, cobalt, iron, manganése, copper and bismuth have been separated by TLC on silica gel D with benzene–chloroform (3:1) [687]. Their inhibitory effect on urease has been used for detection of traces of heavy metals after their TLC separation [720]. Daneels [604] and co-workers applied thin-layer chromatography on silica gel fixed with starch, to separations of rare-earth elements. After the impregnation of the layer with buffers or acids they developed in a solution of di(2-ethylhexyl) hydrogen phosphate in carbon tetrachloride and achieved a few distinct separations, especially of yttrium, lanthanum, cerium, terbium, gadolinium, europium, and samarium. The rapid separation of up to five adjacent rare-earth metals is possible by thin-layer chromatography on silica gel H containing 6% of soluble starch as binder. Massart and co-workers [661] develop the plates by the ascending technique with $0.1M-1.0M$ di(2-ethylhexyl) hydrogen phosphate in carbon tetrachloride. Oguma [662] recommends separating scandium, yttrium, rare-earth metals, thorium and uranium in hydrochloric acid solution by developing on cellulose layers with dioxan–$12M$ hydrochloric acid or dioxan–$14M$ nitric acid mixtures as solvents. Vagina and Volynets [663] described the separation and determination of $0.01-0.1\%$ of iron, copper, mecury, zinc or calcium in rare-earth metal preparations by thin-layer chromatography with silica gel and water or benzene–tributyl phosphate mixtures. The separation of microgram amounts of cerium(III) and cerium(IV) from rare-earth mixtures also gave good results [664].

Separation of selenium, tellurium and chromium on a silica gel layer has been described by Kayashi and Ogata [605]. They use methyl isobutyl ketone, ethyl acetate, butyl acetate, methanol saturated with $2M$ hydrochloric acid, or acetone. The silica gel was first treated with $6M$ hydrochloric acid for $3-4$ hours and then dried at 120° for an hour. The hydrogen sulphide group has been separated by TLC, and solvent mixtures for this purpose have been examined [722]. Arsenic(III) and antimony(III) and tin(II) can also be well separated on thin layers. A suitable solvent is a mixture of 1-butanol and hydrochloric acid (20 : 1) [606]. Senf [665] separates arsenic, antimony and tin as tetramethyldithiocarbamates after their extraction into chloroform, using a silica gel–alumina (2 : 1) thin layer and benzene – chloroform (2 : 1) as solvent. Arsenic and antimony can be separated on silica gel layers with $1M$ tartaric acid-butanol (1 : 1) as solvent [666]. For the separation of arsenic(III) and (V) a mixture of acetone with $15M$ phosphoric acid (50 : 1) and a silica gel layer with 5% calcium sulphate as binder may be used successfully [610]. A layer of silica gel bound with calcium sulphate is also suitable for separations in the platinum metal group. With acetone, hydrochloric acid and ethyl acetate (100 : 1 : 100), or acetone, nitric acid and ethyl acetate (500 : 1 : 500 or 5 : 2 : 5), good separations can be achieved [608]. Trujillo and Frye [667] recommend a mixture of acetone–ethanolamine–water (87 : 3 : 10) as a good solvent for this separation, on 0.275 mm thick silica gel AR layers. Semiquantitative determination of noble metals is possible after separation with acetone–acetylacetone–water–$2M$ hydrochloric acid (20 : 2 : 1 : 1) as solvent, on layers of aluminium oxide, with an error of $\pm 6\%$ [668]. Various solvents have been proposed by Gagliardi and Shambri [721] for separation of the platinum metals on cellulose thin layers. For the separation of gallium and aluminium, the thin-layer method is also successful. On a layer of silica gel with gypsum as binder and with an acetone–hydrochloric acid mixture (200 : 1) as solvent, a very sharp separation can be achieved within $10-15$ minutes, leaving aluminium on the start, while gallium moves with the solvent. TLC on cellulose with 1-butanol–hydrochloric acid–water (8 : 1 : 1) has been used to separate gallium, iron, aluminium, indium and titanium [723]; on silica gel F with 2-butanol–$2M$ hydrochloric acid (11 : 3) for germanium, arsenic and antimony, and with acetone–$3M$ hydrochloric acid (4 : 1) on cellulose for iron, titanium and vanadium [724]. Zirconium and niobium can be separated within 45 minutes, with a mixture of methanol and hydrofluoric acid [187], and titanium, zirconium, thorium, scandium and uranium have been successfully separated with a mixture of butyl phosphate and nitric acid. Oguma [669] separates zirconium from 19 metal ions, including scandium, yttrium, thorium, uranium and rare–earths. Niobium was separated from tantalum by Dragulescu and co-workers [670] on layers of cellulose mixed with Dowex 1, with $0.7M$ hydrochloric acid and $2M$ oxalic acid in acetone and water (1 : 1) as solvent mixture. For the separation of rhenium, tungsten and molybdenum, silica gel was recommended and the best results were obtained by using methanol–$3M$ hydrochloric acid (7 : 3) as solvent [671]. These ions and vanadium can also be separated with butanol–methanol, ethanol–methanol and propanol–methanol (or their mixtures with acetic and hydrochloric acid and water) [672]. Kuroda *et al.* have investigated DEAE-cellulose as stationary phase, with various acid–organic solvent mixtures [737–739].

Some good possibilities are given by thin-layer chromatography for the separation of uranium. Selective separation from 52 other metals, including the lanthanides, platinum metals, alkali and alkaline–earth metals, scandium, titanium, vanadium, tungsten, molybdenum and zirconium can be carried out on layers of silica gel HR or cellulose, with methyl isobutyl ketone–tributyl phosphate–$4M$ nitric acid (100 : 1 : 10; upper phase) as solvent [673]. Separation of uranium and plutonium from transplutonium elements was described by Volynets and Guseva [674]. With silica gel layers and tributyl phosphate–benzene (1 : 1) as solvent, good results were obtained from samples in which the ratio of uranium to polonium was 500 : 1 and that of uranium to americium 20000 : 1. Thin-layer methods are also very convenient for separations in radiochemistry. Seiler [675] separated molybdenum-99 and technetium-99 on cellulose layers with butanol saturated with $1M$ hydrochloric acid; Andreev and co-workers [676] recommended the determination of radium-226 and radium-224 by use of a barium sulphate layer.

Separation of *EDTA* complexes on microcrystalline cellulose has been used for separation of transition metals [741]. For quantitative determination of chromate, Siechowski [609] recommends a 0.5-mm silica gel layer and development with a mixture of methanol and water (4 : 1). The separation takes 6 minutes. The R_F value of chromate is 0.90 and it can be determined planimetrically. Tri-n-alkylamine oxide and tri-n-octylarsine oxide have been used on silica gel thin layers for reversed-phase chromatography of several ions [731]. Various long-chain amines have been similarly investigated [733]. Rare earths, thorium, zirconium and uranium have been separated (the lanthanides in pairs or groups) by TLC on silica gel impregnated with Amberlite LA-2 (sulphate form), with various sulphuric acid–ammonium sulphate solvent systems [732].

The use of hydrofluoric acid solvent systems for reversed-phase TLC on cellulose powder impregnated with long-chain amines was reported on by Brinkman *et al.* [735].

Thin-layers of polytetrafluoroethylene impregnated with di(2-ethylhexyl) hydrogen phosphate have been used to separate a number of metal ions [736].

Thin-layer chromatography is also suitable for complex ions. Druding and Hagel [611] separated the *cis* and *trans* amino-acid complexes of cobalt(III), using a mixture of methanol, dimethylsulphoxide and 70% perchloric acid (80 : 120 : 1). Yamaoka and co-workers [607] used the thin-layer technique for the study of the composition of the dithiophosphate complexes of palladium(II).

For the separation of anions a number of thin-layer methods have been described. Several are suitable for halides. Bark and co-workers [612] recommended cellulose layers and various mixtures of acetone and water as the solvent. R_F values increase with increasing atomic number and are practically independent of the cation, if this is sodium, potassium or ammonium. The authors explain the fact that separation is sharp and the spots compact in terms of the ion-exchange behaviour of cellulose. Good separations of halides [613], and also of other anions such as nitrite, thiosulphate, chromate, azide, cyanide, thiocyanate, borate, sulphide, arsenite, arsenate, nitrate, sulphate and phosphate [614] have been achieved on a layer of maize starch. On a silica gel layer, Takeuchi and Tsunoda [615] separated fluoride, chloride, bromide, iodide, iodate, bromate and chlorate, using a mixture of acetone, ethyl methyl ketone and 14% aqueous ammonia (6 : 4 : 1). R_F values increase in the order given above. Muto [616, 617] proposed precipitation chromatography on thin layers for the separation of halides. He prepared the layer by suspending silica gel in 3% silver nitrate solution and developed with water saturated with isobutyl alcohol and 40% aqueous ammonium acetate solution (4 : 1). The separation is good, and the detection is carried out under ultraviolet light. The chromatogram can be estimated quantitatively directly, because the length of the zone is directly proportional to the concentration of the ions present. In the same way phosphates can also be separated and estimated.

TLC on silica gel impregnated with 5% polyvinyl alcohol has been used to separate molybdate, perrhenate and selenite, the solvent being methanol and water with either hydrochloric or nitric acid. The R_F values are a function of the acid concentration, but least so with the nitric acid media [725]. Three-component mixtures of selenite, tellurite, vanadate and molybdate have been separated on silica gel G with diethyl oxalate–hydrochloric acid (60 : 1) or n-butyl acetate–hydrochloric acid (200 : 3) [726]. Thiourea, thiocyanate and sulphide have been separated by circular TLC with methanol (the best solvent), ethanol, pyridene and 1-butanol–water–pyridine–ammonia (8 : 8 : 4 : : 1) [727].

Appreciable attention has been devoted to the separation of phosphorus compounds on thin layers. On pure powdered cellulose, inorganic acids of phosphorus can be separated by using a mixture of methanol, concentrated ammonia solution and 10% aqueous trichloroacetic acid solution (10 : 3 : 1 : 6) [618]. For the separation of phosphorus oxyacids of various oxidation states the technique of two-dimensional chromatography on cellulose layers can be used [619]. The first development is done with a mixture of methanol, concentrated ammonia solution, water and 25% trichloroacetic acid solution (6 : 7 : 5 : 2), adjusted to pH 5 with ammonia. In the second direction a mixture of methanol, concentrated ammonia solution, water and trichloroacetic acid (55 ml + 5 ml + 40 ml + 3 g) is used for development. There is a relationship between the structure of the separated phosphorus compounds and their R_F values. For the separation of condensed phosphates, cellulose layers may be used, as well as mixed layers composed of cellulose and silica gel, or starch layers. Aurenge and co-workers [620] prepared the layers by spraying an aqueous cellulose suspension onto a glass plate, drying, and developing for twelve hours with the solvent system to be used. The plates are dried for 30 minutes, and are ready for use. For the separation of linear polyphosphates, acidic mixtures are most suitable. The first eight members of the series can be successfully separated in a mixture of water, ethanol, isobutyl alcohol, propanol and concentrated ammonia solution (30 : 35 : 15 : 20 : 0.4), and 5 g of trichloroacetic acid. Cyclic polyphosphates can be separated from linear ones and from each other by a solvent containing concentrated ammonia solution, methanol, isobutyl alcohol and water (9 : 50 : 10 : 31). Canić and co-workers [621] separated condensed phosphates on a starch layer with solvents containing EDTA. For quantitative estimation of the separated condensed phosphates, colorimetry is suitable. It is carried out in the same manner as in paper chromatography, for example by the Rössel and Kreisslich method [622]. The developed chromatogram is immersed in a 6% solution of collodion in 7% glycerol. A chromatogram treated in this way can be analysed in the same way as a paper chromatogram, i.e. the spots can be cut out, mineralized, and the content of phosphorus determined colorimetrically. Also silica gel-sodium polyacrylate or silica gel-cellulose acetate layers are suitable for the separation of condensed phosphates [677]. Several linear and cyclic phosphates were separated by Tanzer and co-workers [678] on plates coated with layers of polyethylene-imine-impregnated microcrystalline cellulose by anion-exchange thin-layer chromatography. Various oxyacids of phosphorus were examined by several thin-layer techniques by Van Ooij and Houtman [679]. Iida and Yamabe [734] used Merck PEI-cellulose F plates to separate various polyphosphates, using sodium or ammonium chloride for development.

Thin-layer chromatography can also be employed for the separation of a series of complex ions. Thus, for example, mercury, copper, cadmium, nickel and zinc can be separated according to Deguchi [623] on silica gel impregnated with 0.3% dithizone in chloroform. Best results are obtained with a solvent system composed of $1-2M$ acetic acid and acetone (1 : 1) or $2-4M$ acetic acid and acetone (10 : 6.6). With circular chromatography the zone closest to the start is orange [mercury(II)], followed by a grey-black zone of copper, an orange one of cadmium, a purple one of nickel, and a pink one of zinc. Metal dithizonates can be separated by extracting them with chloroform, applying the solution to a cellulose layer, and developing with a mixture of hexane and benzene (4 : 1) [624]. Complexes of iron(II) and (III), cobalt, nickel and copper with 1-nitroso-2-naphthol, extracted with chloroform from a medium containing acetic acid (at pH 4) [625] can be separated on a silica gel layer. The first development is made with a mixture of benzene and amyl ether (2 : 2) and when the solvent front has moved 12.5 cm the chromatography is interrupted, the chromatogram dried and a second development made with a mixture of benzene and dioxan (40 : 17). Senf [626] proposed separation of these ions on a layer of silica gel and alumina (1 : 1) with benzene as solvent. The R_F values obtained are Co 0.41, 0.32 (two spots), Ni 0.0, Mn 0.19, Fe^{3+} 0.50, Fe^{2+} 0.0. Senf recommends a silica gel layer for separation of the diethyldithiocarbamates of mercury, lead, copper, bismuth and cadmium, using a solvent system composed of citric acid and phosphate buffer (pH 8) [627].

Dithizone complexes of cadmium, copper, mercury, manganese, nickel, lead, zinc and bismuth were separated on silica gel layers by Takeuchi and Tsunoda [680], using various solvent systems. Muchová and Jakl [681] studied the separation of various ions with diethyldithiocarbamic acid, dithizone and 8-hydroxyquinoline as chelating agents on silica gel layers. Trehan [682] separated copper, nickel and cobalt on silica gel G layers by development with a mixture of cyclohexane, butanol and acetylacetone (84 : 14 : 2), shaken with 10% ammonia solution, before use. Separation of metal-EDTA complexes with thin-layer chromatography was studied by Vanderdeelen [683]. Good separations of cobalt, copper, nickel, manganese, chromium and iron complexes were obtained on silica gel H layers with water–2-methoxyethanol–ethyl methyl ketone–acetone–aqueous ammonia (40 : 20 : 20 : 5 : 15) as solvent. Wasilewska [684] recommended separation of 1-(2-pyridylazo)-2-naphthol complexes of copper, cobalt and iron on layers of silica gel G and diatomaceous earth G (Merck) (1 : 1) with hexane–dioxan–methanol–ethyl methyl ketone–45% formic acid (40 : 20 : 5 : 5 : 1) as solvent. For the separation of some inorganic cobalt complexes Haworth and Zetmeils [685] used microcrystalline cellulose as the separation medium. Separation of inert complexes of cobalt and chromium by thin-layer chromatography on silica gel was described by Baba and co-workers [686].

Reversed-phase chromatography on thin layers has also often been successfully employed. With tributyl phosphate (TBP) as the stationary phase, niobium and tantalum can be separated with a mixture of ammonium thiocyanate, 10% oxalic acid solution and $6M$ hydrochloric acid (1 : 10 : 10), and platinum group elements with $2M$ hydrochloric acid [628]. Bark and co-workers [629] investigated in detail the separations on TBP with varying concentrations of hydrochloric acid, and explained the separation process by the ability of metal ions to form complexes with chloride.

6.3. Ion-Exchange Chromatography

In the introduction it was said that the principles of separation on ion-exchangers would not be discussed, because its importance in inorganic analysis is such that numerous other publications have already been devoted to it. Recently, however, separation techniques on ion-exchangers, similar to paper chromatography, have developed so rapidly that it is necessary to consider ion-exchange chromatography at this point.

The first method to be considered is the use of papers impregnated with substances capable of ion-exchange, both resins and liquid ion-exchangers being suitable. Various papers impregnated with well-known ion-exchange resins, such as Dowex, or Amberlite, etc., are commercially available. These give results comparable with those obtained with ion-exchange column chromatography, if the same solvent systems are used. This was demonstrated, for example, by Lederer and Ossicini [630] for about 30 metal ions, using a system of anion-exchangers with hydrochloric acid as eluent. They also demonstrated that differences exist only in the technical side of the process.

As an example of the group of separations on papers impregnated with commercial ion-exchangers, the separation of arsenic from the majority of other ions may be mentioned. It can be done successfully with $0.05M$ malonic acid, on papers impregnated with Amberlite SA2 [107]. With the same papers [397] or papers impregnated with anion-exchangers, e.g. Whatman DE-20 paper [390], the possibility of separation of rare earths has been investigated. Ossicini [480] proposed Amberlite SB-2 and WB-2 or Whatman weakly alkaline ion-exchange papers for the separation of anions. He achieved a number of sharp separations with potassium nitrate solution. Papers impregnated with commercial exchangers are also suitable for the separation of uranium, copper, thorium and lanthanum [310]; and copper, cobalt, nickel, and cadmium [362]; or uranium from a variety of mixtures [310, 325, 326]. Their use for the separation of technetium(IV), (V) and (VII) is also interesting. Mono-, di- and triethanolamine have been used as solvents for separation of copper, nickel, zinc, cadmium, mercury, lead and calcium on ion-exchange papers [713]. Uranium has been separated from other ions on diethylaminoethyl cellulose anion-exchange paper, with 1-butanol–acetic acid–water (1 : 1 : 1) as solvent [714]. Scandium, zirconium (hafnium), nickel and thorium have been separated on cation-exchange papers by mixed solvents containing organic extractants such as dithizone, dimethylsulphoxide and tri-n-octyl phosphine oxide [715]. Some lanthanides have been separated on Whatman SA-2 paper [730]. The adsorption of chloroauric acid on various exchange papers has been studied [740].

Tri-n-octylamine has been found suitable for a variety of separations. On papers impregnated with it, thorium, uranium and lanthanum [317] can be separated very well. Uranium, lanthanum and samarium can be well separated on such papers with ammonium nitrate solution at pH 2.4 [633]. Waksmundski and Przeszlakowski [634] investigated the behaviour of 33 metals on papers impregnated with tri-n-octylamine, using $0.5 - 10M$ hydrochloric acid, or $0.5 - 4M$ sodium chloride as the developing solvent. They found low R_F values for certain cations (Zn^{2+}, Cd^{2+}, Hg^{2+}, Sn^{4+}, Au^{2+}, Pt^{4+}), and high ones for others (Al^{3+}, Be^{2+}, Ni^{2+}, Cr^{3+}, Th^{4+}, La^{3+}, Sc^{3+}, Y^{3+}, Ce^{3+}). They also defined a group of cations having R_F values which depend considerably on the concentration of chloride ions in the solvent. Knoch and Lahr [635] saturated the nitrates of plutonium(III), americium(III), and curium(III), obtaining a very good separation, using $3M$ lithium nitrate in $0.02M$ nitric acid.

Werner [636] proposed the impregnation of the paper with dinonylnaphthalenesulphonic acid. Using as solvent a lactic acid solution adjusted to a suitable pH with ammonia, he was able to carry out sharp separations within the lanthanide group. The lanthanides can also be separated on papers impregnated with di(2-ethylhexyl)phosphoric acid [389, 391] acidified with hydrochloric acid or $0.1M$ oxalic acid or acidified sodium sulphate. Cerrai and Ghersini investigated the behaviour of 57 cations on such papers. For chromatography they used 10^{-4}–$10M$ hydrochloric acid as eluent [637].

Suitable conditions were found for the separation of the alkaline earth metals and for their separation from other groups of elements [198] as well as a sharp separation of aluminium from indium, thallium and especially gallium [284].

O'Laughlin and co-workers impregnated papers for the separation of lanthanides, thorium, and uranium with methylene bis[(n-hexyl)phosphine oxide] [638] and methylene bis [(2-ethylhexyl)phosphine oxide] [640]. For the separation of selenium and tellurium it is advantageous to impregnate the paper with tributyl phosphate. As a suitable mobile phase, $1-2M$ hydrobromic acid or $8M$ sodium bromide [639], or $1-5M$ nitric acid have been recommended. When $2-6M$ hydrochloric acid is used for development, gold, silver, and platinum group metals can also be separated distinctly [641]. Lyle and Nair [746] used papers impregnated with di(2-ethylhexyl) hydrogen phosphate and 2-ethylhexyl dihydrogen phosphate for separation of lanthanides, strontium, barium, yttrium, manganese, iron, cobalt, nickel, gold, platinum, palladium and silver.

For certain separations impregnation of papers with various inorganic compounds has been found practical. On papers impregnated with zirconium phosphate, arsenites and arsenates can be separated very well by $1M$ hydrochloric or nitric acid. With $0.1M$ hydrochloric acid it is possible to distinctly separate copper (R_F $0.7-0.8$) from silver, which remains on the start. Further development of the chromatogram with a mixture of ammonia and ammonium chloride makes silver move. A simple separation of silver from lead can be achieved by using $0.1M$ sulphuric acid for development. Lead remains near the start (R_F 0.17) while silver and mercury move closely behind the solvent front [93]. Lederer and co-workers [310] investigated the possibility of separating uranium, copper, thorium and lanthanum on papers impregnated with zirconium phosphate. On paper impregnated with zirconium molybdate [181] calcium, strontium and barium can be separated with $0.5M$ hydrochloric acid, in which they move with the solvent front. Separation from each other can be achieved on the same type of paper with $0.9M$ ammonium chloride. The possibility of using hydrochloric or perchloric acid for separations on papers impregnated with stannic phosphate has been investigated in detail [642].

Qureshi and co-workers have examined papers impregnated with several types of inorganic compound, notably titanium tungstate [743], tin(IV) tungstate [744, 745] and tin(IV) selenite [745].

Thin layers are also sometimes suitable. Cellulose or some of its derivatives [643], or suitably impregnated silica gel are used for the preparation of the layers. Holzapfel and co-workers [644, 645] impregnated silica gel with di(2-ethylhexyl)phosphoric acid and succeeded in separating all pairs of the following elements by the thin-layer technique: La, Ce, Pr, Nd, Sm, Eu, Gd, Tb, Dy, Ho, Er, Y, and Yb, with the exception of the pairs Pr-Nd and Er-Y. They used hydrochloric acid or nitric acid of suitable concentration for the development. Better results were obtained with nitric acid.

REFERENCES

1. SCHWAB, G. M. and JOCKERS, K. *Naturwiss.* **25**, 44 (1937).
2. SCHWAB, G. M. and JOCKERS, K. *Z. Angew. Chem.* **50**, 546 (1937).
3. SCHWAB, G. M. and DATTLER, G. *Z. Angew. Chem.* **50**, 691 (1937).
4. SCHWAB, G. M. and DATTLER, G. *Z. Angew. Chem.* **51**, 709 (1938).
5. SCHWAB, G. M. and GHOSH, A. N. *Z. Angew. Chem.* **52**, 666 (1939).
6. SCHWAB, G. M. and GHOSH, A. N. *Z. Angew. Chem.* **53**, 39 (1940).
7. VENTURELLO, G. *Atti. Acad. Sci. Torino, Classe Sci. Fis. Mat. e. Nat.*, **79**, 288 (1944).
8. FLOOD, H. *Z. Anal. Chem.* **120**, 327 (1960).
9. JACOBS, P. W. M. and TOMPKINS, F. C. *Trans. Faraday Soc.* **41**, 388 (1945).
10. NEUGEBAUER, W., and SCHÄFER, H. *Z. Anorg. Allgem. Chem.* **273**, 227 (1953).
11. SCHWAB, G. M. and ISSIDORIDIS, A. *Z. Phys. Chem.* (*Leipzig*), *B* **53**, 1 (1942).
12. SIEWERT, G. and JUNGNICKEL, H. *Z. Anorg. Chem.* **257**, 215 (1948).
13. OLSHANOVA, K. M. and CHMUTOV, K. V. *Zh. Analit. Khim.* **11**, 94 (1956).
14. PINTEROVIĆ, Z. *Bull. Soc. Chem. Belge* **58**, 522 (1949).
15. KERŠULIN, M. and ŠVARC, Z. *Kem. Vjestnik* (*Zagreb*) **17**, 99 (1943).
16. NEDOBORA, A. F. *Ref. Zh. Khim.* **52**, 162 (1955).
17. OLSHANOVA, K. M. and CHMUTOV, K. V. *Zh. Analit. Khim.* **13**, 162 (1958).
18. VIKHUTINSKIĬ, A. A. *Sbornik Stud. Nauch. Robot. Moskov. Inst. Narod. Khoz.* 26 (1956).
19. TANAKA, M., ASHIZAWA, T. and SHIKATA, M. *Chem. Research* (*Japan*) **5**, *Inorg. Anal. Chem.* 35 (1949).
20. PINTEROVIĆ, Z. *Kem. Vjestnik* (*Zagreb*) **15/16**, 45 (1941).
21. LISTER, B. A. J. *Z. Appl. Chem.* **2**, 280 (1952).
22. GURVICH, A. M., GAPON, T. B. and RABINOVICH, M. C. *Khim. Promyshlennost* 31 (1956).
23. FRICKE, R. and SCHMÄH, H.: *Z. Anorg. Allgem. Chem.* **255**, 253 (1948).
24. BEAUCOURT, J. H. and MASTERS, D. L. *Metallurgia* **32**, 181 (1945).
25. BACH, R. O. and GARMENDIA, A. A. *An. Asoc. Quím. Argent.* **39**, 191 (1951).
26. DEAN, J. A. *Anal. Chem.* **23**, 1096 (1951).
27. VENTURELLO, G. *Ric. Sci. Progr. Tech.* **14**, 256 (1943).
28. SCHWAB, G. M. and GHOSH, A. N. *Z. Anorg. Chem.* **258**, 323 (1949).
29. ERÄMETSÄ, O. *Bull. Commis. Geol. Finlande* **14**, No. 126, 36 (1941).
29a. ERÄMETSÄ, O. *Suomen Kemistilehti* **16B**, 13 (1943).
30. LINDNER, R. and PETER, O. *Z. Naturforsch.* **1**, 67 (1946).
31. CROATTO, N. *Ric. Sci. Progr. Tech.* **12**, 1197 (1941).
32. MARTIN, M., JENG-TSONG, Y. and DAUDEL, P. *Anal. Chim. Acta* **3**, 222 (1949).
33. NYDAHL, F. *Anal. Chem.* **26**, 580 (1954).
34. BACH, R. O. *An. Asoc. Quím. Argent.* **37**, 55 (1949).
35. GAPON, E. N. and GAPON, T. B. *Dokl. Akad. Nauk SSSR* **60**, 817 (1948).
36. KRAUS, K. A. and PHILLIPS, H. O. *J. Am. Chem. Soc.* **78**, 249 (1956).
37. KRAUS, K. A. and PHILLIPS, H. O. *J. Am. Chem. Soc.* **78**, 694 (1956).
38. KRAUS, K. A., CARLSON, T. A. and JOHNSON, J. S. *Nature* **177**, 1128 (1956).
39. SEN, B. N. *Anal. Chim. Acta* **12**, 154 (1955).
40. SEN, B. N. and RAY, K. *J. Sci. Ind. Res.* (*India*) *Sect. B* **14**, 604 (1955).
41. SEN, B. N. *Anal. Chim. Acta* **23**, 152 (1960).
42. SEN, B. N. *Z. Anorg. Allgem. Chem.* **276**, 208 (1954).
43. GLEMSER, O. and RIEK, G. *Angew. Chem.* **69**, 91 (1957).
44. MARJANOVIĆ-KRAJOVAN, V., TURINA, S., INDJIĆ, N. and JOVANOVIĆ, V. *Z. Anal. Chem.* **195**, 427 (1963).
45. HANSEN, R. and GUNNAR, K. *J. Am. Chem. Soc.* **71**, 4158 (1949).
46. KUTEINIKOV, A. F. and BRODSKAYA, V. M. *Zh. Analit. Khim.* **17**, 305 (1962).
47. MARTYNENKO, L. I., EREMIN, G. K. and KAMENEV, A. I. *Zh. Neorgan. Khim.* **4**, 2639 (1959).
48. SHEMYAKIN, F. M. and MITSELOVSKII, Z. S. *Zavodsk. Lab.* **16**, 748 (1950).
49. BOZHEVOLNOV, E. A. and KARAKOVSKAYA, O. A. *Zavodsk. Lab.* **27**, 11 (1961).
50. VYDRA, F. and MARKOVÁ, V. *Talanta* **10**, 711 (1963).
51. PIERCE, T. B. and PECK, P. F. *Anal. Chim. Acta* **26**, 557 (1962).
52. HILLAIRD, L. B. and FREISER, H. *Anal. Chem.* **24**, 752 (1952).
53. KRAUSZ, I. *Magyar Kém. Foly.* **66**, 218 (1960).

54. ŠULCEK, Z., MICHAL, J. and DOLEŽAL, J. *Collection Czech. Chem. Commun.* **24**, 1815 (1959).
55. ŠULCEK, Z., MICHAL, J. and DOLEŽAL, J. *Chemist-Analyst* **50**, 13 (1961).
56. ŠULCEK, Z., MICHAL, J. and DOLEŽAL, J. *Collection Czech. Chem. Commun.* **26**, 246 (1961).
57. ŘEZÁČ, Z. and ROUBAL, M. *Collection Czech. Chem. Commun.* **23**, 426 (1958).
58. ŠULCEK, Z., DOLEŽAL, J., MICHAL, J. and SYCHRA, V. *Talanta*, **10**, 3 (1963).
59. ERLENMEYER, H. and DAHN, H. *Helv. Chim. Acta* **22**, 1369 (1939).
60. ROBINSON, G. *Metallurgia* **37**, 45 (1947).
61. ERLENMEYER, H. and SCHOENAUER, W. *Helv. Chim. Acta* **24**, 878 (1941).
62. ERLENMEYER, H. and SCHMIDLIN, J. *Helv. Chim. Acta* **24**, 1213 (1941).
63. GURVICH, A. M. *Zh. Analit. Khim.* **11**, 437 (1956).
64. BURRIEL-MARTI, F. and PINO PEREZ, F. *Anal. Chim. Acta* **3**, 468 (1949).
65. HOPF, P. P. *J. Chem. Soc.* 785 (1946).
66. SHEMYAKIN, F. M. *Zh. Analit. Khim.* **3**, 359 (1948).
67. BACH, R. O. *Ind. y Quím. (Buenos Aires)* **12**, 283 (1950).
68. PICKERING, W. F. *J. Chromatog.* **1**, 274 (1958).
69. PICKERING, W. F. *J. Chromatog.* **4**, 477 (1960).
70. PICKERING, W. F. *J. Chromatog.* **4**, 481 (1960).
71. PICKERING, W. F. *J. Chromatog.* **4**, 485 (1960).
72. BECKMANN, T. J. and LEDERER, M. *Congrès Nucleaire, Rome,* July 1960; *Chem. Abstr.* **58**, 1896b (1963).
73. PLUCHET, E. and LEDERER, M. *J. Chromatog.* **3**, 290 (1960).
74. BECKMANN, T. J. and LEDERER, M. *J. Chromatog.* **3**, 498 (1960).
75. LEDERER, M. *J. Chromatog.* **4**, 414 (1960).
76. BECKMANN, T. J. and LEDERER, M. *J. Chromatog.* **5**, 341 (1961).
77. LEDERER, M. and SHUKLA, S. K. *J. Chromatog.* **6**, 353 (1961).
78. LEDERER, M. *J. Chromatog.* **6**, 437 (1961).
79. LEDERER, M. *J. Chromatog.* **6**, 518 (1961).
80. LEDERER, M. *J. Chromatog.* **7**, 366 (1962).
81. MURTHY, A. R. V., NARAYAN, V. A. and RAO, M. R. A. *Current. Sci. (India)* **24**, 158 (1955).
82. PFEIL, E., PLOSS, G. and SARAN, H. *Z. Anal. Chem.* **146**, 241 (1955).
83. TEWARI, S. N. *Z. Anal. Chem.* **141**, 401 (1954).
84. WEISS, A. and FALLAB, S. *Helv. Chim. Acta* **37**, 1253 (1954).
85. RAO, V. K. M. *J. Sci. Ind. Res. (India)* **19B**, 171 (1960).
86. BARNABAS, T. and BARNABAS, J. *Naturwiss.* **42**, 15 (1955).
87. LEDERER, M. *Nature* **162**, 776 (1948).
88. MUTO, S. *J. Chem. Soc. Japan, Pure Chem. Sect.* **76**, 284 (1955).
89. MAJUMDAR, A. K. and CHAKRABARTTY, M. M. *Anal. Chim. Acta* **17**, 415 (1957).
90. MAJUMDAR, A. K. and CHAKRABARTTY, M. M. *Anal. Chim. Acta* **19**, 132 (1958).
91. REEVES, W. A. and CRUMPLER, T. B. *Anal. Chem.* **23**, 1576 (1951).
92. WARREN, G. W. and FINK, R. W. *J. Inorg. Nucl. Chem.* **2**, 176 (1956).
93. NUNES DA COSTA, M. J. and JERONIMO, M. A. S. *J. Chromatog.* **5**, 456 (1961).
94. BERGAMINI, C. and VERSORESE, W. *Anal. Chim. Acta* **10**, 328 (1954).
95. POLLARD, F. H., McOMIE, J. F. W. and ELBEIH, I. I. M. *J. Chem. Soc.* 466 (1951).
96. KIELCZEWSKI, W. and TOMKOWIAK, J. *Chem. Analit. (Warsaw)* **7**, 925 (1962).
97. JANJIĆ, T. J., ĆELAP, M. B. and ŠPANOVIĆ, Ž. F. *Bull. soc. chim. (Belgrade),* **25/26**, 531 (1960—61).
98. SURAK, J. G. and MARTINOVICH, R. J. *J. Chem. Educ.* **32**, 95 (1955).
99. AGRINIER, H. *Compt. Rend.* **257**, 145 (1963).
100. KOLIER, I. and RIBAUDO, C. H. *Anal. Chem.* **26**, 1546 (1954).
101. LEDERER, M. *Anal. Chim. Acta* **3**, 476 (1949).
102. PEG, V. and PARELLADA, L. *Clin. y Lab.* **63**, 263 (1957).
103. RAO, V. K. M. *J. Sci. Ind. Res. (India)* **18B**, 492 (1959).
104. TRIPATHI, D. N. and TEWARI, S. N. *J. Prakt. Chem.* **9**, 1 (1959).
105. D'AMORE, G. *Ann. Chim. (Roma)* **46**, 517 (1956).
106. SASTRI, M. M. and RAO, A. P. *J. Chromatog.* **9**, 250 (1962).
107. SHERMA, J. and CLINE, C. W. *Talanta* **10**, 787 (1963).
108. AGRINIER, H. *Commisariat à l'énergie atomique. Rapport CEA* 2099.
109. DILLER, H. and REX, O. *Z. Anal. Chem.* **137**, 241 (1952).

110. POLLARD, F. H., McOMIE, J. F. W. and BANISTER, A. *J. Chem. Ind. (London)* **49**, 1598 (1955).
111. PFEIL, E. and GOLDBACH, H. J. *Arch. Toxikol.* **16**, 134 (1956).
112. TEWARI, S. N. *Z. Anal. Chem.* **180**, 109 (1961).
113. VAECK, S. V. *Anal. Chim. Acta* **10**, 48 (1954).
114. HEYNDRYCKX, P. *Publ. Groupe Avan. Méthodes Spectrog.* 373 (1960).
115. MURATA, A. *J. Chem. Soc. Japan, Pure Chem. Sect.* **78**, 520 (1957).
116. THILO, E., SCHULZ, G. and WICHMAN, E. M. *Z. Anorg. Allgem. Chem.* **272**, 182 (1953).
117. THILO, E. and RATZ, R. *Z. Anorg. Allgem. Chem.* **260**, 255 (1949).
118. JANJIĆ, T. J. and ĆELAP, M. B. *Bull. soc. chim. (Belgrade)* **25/26**, 133 (1960—61).
119. LABICH, YU. M. *Ukr. Khim. Zh.* **28**, 641 (1962).
120. BURSTALL, F. H., DAVIES, G. R., LINSTEAD, R. P. and WELLS, R. A. *J. Chem. Soc.* 516 (1950).
121. ŠTEFFEK, M. *Chem. Listy* **56**, 951 (1962).
122. LEDERER, M. *Anal. Chim. Acta* **2**, 261 (1948).
123. HARASAWA, S. *J. Chem. Soc. Japan, Pure Chem. Sect.* **72**, 423 (1951).
124. ELBEIH, I. I. M., McOMIE, J. F. W. and POLLARD, F. H. *Disc. Faraday Soc.* **7**, 183 (1949).
125. ANDERSON, J. R. and MARTIN, E. C. *Anal. Chim. Acta* **13**, 253 (1955).
126. VERNOIS, J. *J. Chromatog.* **1**. 52 (1958).
127. BRUNINX, E., EECKHOUT, J. and GILLIS, J. *Mikrochim. Acta* 689 (1957).
128. BLASIUS, E. and CZEKAY, A. *Actas de Congresso Lisboa*, Vol. III, p. 37; *Anal. Chem.* **156**, 81 (1957).
129. SCOTT, I. A. P. and MAGEE, R. J. *Talanta* **1**, 329 (1958).
130. HUNT, E. C., NORTH, A. A. and WELLS, R. A. *Analyst* **80**, 172 (1955).
131. MARTIN, I. and MAGEE, R. J. *Talanta* **10**, 1119 (1963).
132. WILLIAMS, A. R. *J. Chem. Soc.* 3155 (1952).
133. BURSTALL, F. H. and WILLIAMS, A. F. *Analyst* **77**, 983 (1952).
134. BURSTALL, F. H., SWAIN, P., WILLIAMS, A. F. and WOOD, G. A. *J. Chem. Soc.* 1497 (1952).
135. MERCER, R. A. and WILLIAMS, A. F. *J. Chem. Soc.* 3399 (1952).
136. MERCER, R. A. and WELLS, R. A. *Analyst* **79**, 339 (1954).
137. AGRINIER, H. *Bull. Soc. Fr. Minéral.* **80**, 275 (1957).
138. CABRAL PEIXOTO, I. M. *Rev. Port. Quím.* **2**, 51 (1959).
139. MOGHISSI, A. and EDGÜER, E. *J. Chromatog.* **11**, 389 (1963).
140. HUNT, E. C. and WELLS, R. A. *Analyst* **79**, 351 (1954).
141. SOMMER, G. *Z. Anal. Chem.* **151**, 338 (1956).
142. HARTKAMP, H. and SPECKER, H. *Z. Anal. Chem.* **152**, 107 (1956).
143. GORDON, H. T. and HEWEL, C. A. *Anal. Chem.* **27**, 1471 (1955).
144. FILATOVA, A. S. *Vopr. Gigieny Prof. Patol. i Prom. Toksikol. Sverdlovsk, Sb.* 1959, No. 5, 353.
145. ERLENMEYER, H., VON HAHN, H. and SORKIN, E. *Helv. Chim. Acta* **34**, 1419 (1951).
146. DEMETRESCU, A. and CIOACA, E. *Met. Constructia Masini* **12**, 1018 (1960).
147. MAJUMDAR, A. K. and PAL, B. K. *Z. Anal. Chem.* **174**, 429 (1960).
148. MICHAL, J. *Chem. Listy* **50**, 542 (1956); *Collection Czech. Chem. Commun.* **21**, 1295 (1956).
149. DE CARVALHO, R. G. and LEDERER, M. *Anal. Chim. Acta* **13**, 437 (1955).
150. MAGEE, R. J. and HEADRIDGE, J. B. *Analyst* **80**, 785 (1955).
151. RAO, C. L. and SHANKAR, J. *Anal. Chim. Acta* **8**, 491 (1953).
152. OSBORN, G. H. and JEWSBURY, A. *Nature* **164**, 443 (1949).
153. ADER, D. and ALON, A. *Analyst* **86**, 125 (1961).
154. AGRINIER, H. *Chim. Anal. (Paris)* **42**, 600 (1960).
155. LEDERER, M. *Anal. Chim. Acta* **5**, 185 (1951).
156. DE ANDRADE DIAS, B. and PINTO COELHO, F. *Rev. Fac. a. Univ. Coimbra* **26**, 64 (1957).
157. KRAUSZ, I. and ORBÁN, M. *Magyar Kem. Foly,* **69**, 493 (1963).
158. MILES, T. D., DELASANTA, A. C. and BARRY, J. C. *Anal. Chem.* **33**, 685 (1961).
159. BURRIEL-MARTÍ, F. and ANDREU, R. G. *Mikrochim. Acta* 912 (1960).
160. SETTY, T. H. V. *Current. Sci. (India)* **25**, 218 (1956).
161. VERMA, M. R. and GUPTA, P. K. *Current. Sci. (India)* **31**, 10 (1962).
162. HERMANOVICZ, V. and SIKOROWSKA, C. *Gaz. Woda Techn. Sanit.* **29**, 360 (1955).

163. LAMM, C. G. *Acta Chem. Scand.* **7**, 1420 (1953).
164. POLLARD, F. H., BANISTER, A. J., GEARY, W. J. and NICKLESS, G. *J. Chromatog.* **2**, 372 (1959).
165. VENTURELLO, C. and GHE, A. M. *Anal. Chim. Acta* **10**, 335 (1954).
166. PICKERING, W. F. *Anal. Chim. Acta* **8**, 344 (1953).
167. NAGAI, H. and DEGUCHI, T. *Japan Analyst* **12**, 552 (1963).
168. SHUKLA, S. K. *J. Chromatog.* **11**, 93 (1963).
169. IMAI, H. *J. Chem. Soc. Japan Pure Chem. Sect.* **82**, 200 (1961).
170. MINCZEWSKI, J. and FOLDZINSKA, A. *Chem. Analit.* (*Warsaw*) **5**, 575 (1960).
171. LEWANDOWSKI, A. and KIELCZEWSKI, W. *Roczniki Chem.* **30**, 275 (1956).
172. LACOURT, A. and HEYNDRYCKX, P. *Mikrochem. J.* **3**, 181 (1959).
173. LACOURT, A. *Compt. Rend.* **245**, 177 (1958).
174. HERMANOWICZ, W. and SIKOROWSKA, C. *Przemysl Chem.* **8**, 238 (1952).
175. FRIERSON, W. J., REARICK, D. A. and YOE, J. H. *Anal. Chem.* **30**, 468 (1958).
176. BOVALINI, E. and PIAZZI, M. *Ann. Chim.* (*Roma*) **49**, 1075 (1959).
177. LACOURT, A. and HEYNDRYCKX, P. *Mikrochim. Acta* **11**, 1685 (1956).
178. ARNIKAR, H. J. and MEHTA, O. P. *Current Sci.* (*India*) **28**, 400 (1959).
179. GETOFF, N. *Chem. Ztg.* **62**, 334 (1961).
180. KIBA, T., OHASHI, S. and TADA, S. *Bull. Soc. Chem. Japan* **29**, 745 (1956).
181. PEIXOTO CABRAL, J. M. *J. Chromatog.* **4**, 86 (1960).
182. STAMM, H. H. and KOHLSCHÜTTER, R. *Z. Anorg. Allgem. Chem.* **320**, 112 (1963).
183. LEVI, M. C. and DANON, J. *J. Chromatog.* **6**, 269 (1961).
184. DÜSING, W. and KUNZE, H. *Z. Anal. Chem.* **162**, 345 (1958).
185. IMAI, H. *J. Chem. Soc. Japan Pure Chem. Sect.* **82**, 1176 (1961).
186. STEIN, P. C. *Anal. Chem.* **34**, 352 (1962).
187. BRECCIA, A. and SPALLETTI, F. *Nature,* **198**, 756 (1963).
188. NĂSCUȚIU, T. *Acad. Rp. Populare Romîne, Studii Cercetări Chim.* **8**, 653 (1960).
189. ELBEIH, I. I. M. and ABOU-ELNAGA, M. A. *Chemist-Analyst* **47**, 35 (1958).
190. MODREANU, F. *Naturwiss,* **45**, 310 (1958).
191. POLLARD, F. H. and MARTIN, J. V. *Analyst* **81**, 348 (1956).
192. NAKANO, A. and SHIMADA, S. *J. Chem. Soc. Japan, Pure Chem. Sect.* **77**, 836 (1956).
193. GARCIA, A. and AGUIRRE, F. *Scientia* (*Valparaiso, Chile*) **28**, 113 (1961).
194. NAGAI, H. *Kumamoto J. Sci. Ser. A*3, 176 (1957); *J. Chem. Soc. Japan Pure Chem. Sect.* **78**, 840 (1957).
195. TRISTRAM, D. R. and PHILLIPS, C. S. G. *J. Chem. Soc.* 580 (1955).
196. SEILER, H. and ROTHWEILER, W. *Helv. Chim. Acta* **44**, 941 (1961).
197. SOBCZYNSKA, J. and WAWRYCZEK, W. *Milchwissenschaft* **14**, 528 (1959).
198. CERRAI, E. and GHERSINI, G. *J. Chromatog.* **15**, 236 (1964).
199. BARCZA, L. and MESTER, O. *Ann. Chim. Univ. Sci. Budapest, Rolando Eotvos Nominatae, Sect. Chim.* **4**, 33 (1962).
200. SEILER, H., SORKIN, E. and ERLENMEYER, H. *Helv. Chim. Acta* **35**, 120 (1952).
201. SEILER, H. *Helv. Chim. Acta* **46**, 2629 (1963).
202. VENDER, M. *Chem. Listy* **49**, 771 (1955).
203. SOÓS, P. and BLAZSEK, A. *Lucrarile Prezentate Conf. Natl. Farm., Bucharest* 1958, p. 120.
204. ĆELAP, M. B. *Bull. Soc. Chim.* (*Belgrade*) **21**, 225 (1956).
205. HARTKAMP, H. and SPECKER, H. *Z. Anal. Chem.* **158**, 92 (1957).
206. POLLARD, F. H., PITT, E. E. H. and NICKLESS, G. *J. Chromatog.* **7**, 248 (1962).
207. MORETTE, A. and DJAVAD-NIA, D. *Ann. Pharm. Franc.* **20**, 27 (1962).
208. BARNABAS, T., BADVE, M. G. and BARNABAS, J. *Anal. Chim. Acta* **12**, 542 (1955).
209. MODREANU, F., FIŞEL, S. and CARPOV, A. *Naturwiss.* **44**, 615 (1957).
210. POLLARD, F. H., MCOMIE, J. F. W. and MARTIN, J. V. *Analyst* **81**, 353 (1956).
211. DE VRIES, G., HARDONCK, M. J. and BUSCHOW, K. H. J. *Anal. Chim. Acta* **21**, 568 (1959).
212. POLLARD, F. H., MCOMIE, J. F. W. and ELBEIH, I. I. M. *J. Chem. Soc.* 470 (1951).
213. PFEIL, E. *Chemie für Labor u. Betrieb* **5**, 177 (1957).
214. ARNIKAR, H. J. and CHEMLA, M. *Compt. Rend.* **244**, 68 (1957).
215. ALBERTI, G. and GRASSINI, G. *J. Chromatog.* **4**, 423 (1960).
216. BROOMHEAD, J. A. and GIBSON, N. A. *Anal. Chim. Acta* **24**, 446 (1961).
217. MURTHY, A. R. V. and NARAYAN, V. A. *J. Indian Inst. Sci.* **42**, No. 3, 51 (1960).
218. ARNIKAR, H. J. *Nature* **182**, 1230 (1958).

219. BURMA, D. P. *Anal. Chim. Acta* **9**, 513 (1953).
220. BARNABAS, T., BADVE, M. G. and BARNABAS, J. *Naturwiss.* **41**, 478 (1954).
221. FOUARGE, J. and DUYCKAERTS, G. *Anal. Chim. Acta* **14**, 527 (1956).
222. LEDERER, M. *Mikrochim. Acta* **43** (1956).
223. LEDERER, M. *Anal. Chim. Acta* **11**, 528 (1954).
224. MODREANU, F. *J. Chromatog.* **1**, 554 (1958).
225. MARTIN, E. C. *Anal. Chim. Acta* **22**, 142 (1960).
226. VAVRUCH, J., HEJTMÁNEK, M. and BOUZKOVÁ, J. *Chem. Listy* **49**, 1782 (1955).
227. BAYKUT, F. and KÜCÜKUYSAL, A. *Istanbul Univ. Fen. Fac. Mec.*, *Ser. C* **27**, 26 (1962); *Anal. Abstr.* **10**, 2599 (1963).
228. BOJCHINOVA, E. S. and ALESKOVSKII, V. B. *Tr. Leningr. Technol. Inst. Lensoveta* **94** (1958).
229. SMOCZKIEWICZOWA, A. and MIZGALSKI, W. *Chem. Analit.* (*Warsaw*) **4**, 219 (1959); *Nature* **182**, 53 (1958).
230. YAMADA, S. *J. Chem. Soc. Japan Pure Chem. Sect.* **80**, 1444 (1959).
231. LACOURT, A. and HEYNDRYCKX, P. *Compt. Rend.* **241**, 54 (1955).
232. LACOURT, A. and HEYNDRYCKX, P. *Mikrochim. Acta* 1621 (1956).
233. POLLARD, F. H. McOMIE, J. F. W. and NICKLESS, G. *J. Chromatog.* **2**, 284 (1959).
234. ACKERMANN, G. and SCHLOSSER, G. *Z. Chem.* **3**, 471 (1963); **5**, 66 (1965).
235. BERG, E. W. and McINTYRE, R. T. *Anal. Chem.* **27**, 195 (1955).
236. RAO, V. K. M. *J. Sci. Ind. Res, India B*, **20**, 109 (1961).
237. ANDERSON, J. R. A. and LEDERER, M. *Anal. Chim. Acta* **5**, 396 (1951).
238. BERG, E. W. and STRASSNER, J. E. *Anal. Chem.* **27**, 127 (1955).
239. VON HAHN, H., SORKIN, E. and ERLENMEYER, H. *Experimentia* **7**, 258 (1951).
240. WEISS, A., FALLAB, S. and ERLENMEYER, H. *Helv. Chim. Acta* **35**, 1588 (1952).
241. MAGEE, R. J. and HEADRIDGE, J. B. *Analyst* **82**, 95 (1957).
242. TEWARI, S. N. *Kolloid. Z.* **138**, 178 (1954).
243. SINGH, E. J. and DEY, A. K. *Proc. Nat. Acad. Sci. India, Ser. A* **28**, 4 (1959).
244. SINGH, E. J. and DEY, A. K. *J. Indian Chem. Soc.* **37**, 235 (1960).
245. SINGH, E. J. and DEY, A. K. *J. Indian Chem. Soc.* **37**, 763 (1960).
246. SINGH, E. J. and DEY, A. K. *J, Indian Chem. Soc.* **38**, 192 (1961).
247. FIŞEL, S. and GIURGIU, D. *Acad. Rep. Populare Romine, Filiala Iasi, Stud. Cercetari Stiint.*, **11**, 181 (1960).
248. BERG, E. W. and STRASSNER, J. E. *Anal. Chem.* **27**, 1131 (1955).
249. NAGAI, H. *J. Chem. Soc. Japan Pure Chem. Sect.* **76**, 1246 (1956).
250. NAGAI, H. and MUTO, M. *J. Chem. Soc. Japan Pure Chem. Sect.* **82**, 694 (1961).
251. NAGAI, H. *J. Chem. Soc. Japan Pure Chem. Sect.* **78**, 285 (1957).
252. NAGAI, H. and MUTO, M. *J. Chem. Soc. Japan Pure Chem. Sect.* **82**, 694 (1961).
253. NAGAI, H. *J. Chem. Soc. Japan Pure Chem. Sect.* **78**, 840 (1957).
254. NAGAI, H. and KIKUCHI, T. *J. Chem. Soc. Japan Pure Chem. Sect.* **81**, 1271 (1960).
255. NAGAI, H. *J. Chem. Soc. Japan Pure Chem. Sect.* **80**, 617 (1959).
256. NAGAI, H. *Kumamoto J. Sci.*, *Ser. A* **4**, 265 (1960).
257. NAGAI, H. *Japan Analyst* **11**, 95 (1962).
258. YOKOTA, S. and SHIMADA, J. *J. Chem. Soc. Japan Pure Chem. Sect.* **79**, 1144 (1958).
259. INGLE, R. B. and MINSHALL, E. *J. Chromatog.* **8**, 386 (1962).
260. VESELÝ, F., ŠMIROUS, F. and VEPŘEK-ŠIŠKA, J. *Chem. Listy* **49**, 1661 (1955).
261. SCHNEER, A. and ÖRDÖGH, M. *J. Chromatog.* **4**, 319 (1960).
262. LEVI, M. C. and DANON, J. *J. Chromatog.* **3**, 584 (1960).
263. CROUTHAMEL, C. E. and GATROUSIS, C. *Talanta* **1**, 39 (1958).
264. WEATHERLEY, E. G. *Analyst* **81**, 404 (1956).
265. ALMÁSSY, G. and NAGY, Z. *Acta Chim. Hung.* **7**, 325 (1955).
266. TESTA, C. *J. Chromatog.* **5**, 236 (1961).
267. BALAKRISHNAMURTHY, V. V. *J. Sci. Ind. Res., India B*, **20**, 453 (1961).
268. SHIH-FU-TZOU and PEI-LI-TAO *Acta Chim. Sinica* **28**, 344 (1962).
269. CRAWLEY, R. H. A. *Nature* **197**, 377 (1963).
270. BEYER, G. H., JACOBS, A. and MASTELLER, R. D. *J. Am. Chem. Soc.* **74**, 825 (1952).
271. ELBEIH, I. I. M. and GABRA, G. G. *Chemist-Analyst* **52**, 13 (1963).
272. DATTA, S. K. and GHOSE, P. *Z. Anal. Chem.* **158**, 347 (1957).
273. LEDERER, M. *Anal. Chim. Acta* **11**, 132 (1954).
274. LADENBAUER, I. M., BRADACS, L. K. and HECHT, F. *Mikrochim. Acta* 388 (1954).

275. LADENBAUER, I. M. and HECHT, F. *Mikrochim. Acta* 397 (1954).
276. NAGY, Z. and PÓLYIK, E. *Magyar Kém. Foly.* **61**, 248 (1955).
277. LADENBAUER, I. M. and SLAMA, O. *Mikrochim. Acta* 903 (1955).
278. LADENBAUER, I. M. *Mikrochim. Acta* 139 (1955).
279. BERTORELLE, E. and FANFANI, G. *Chim. e Ind.* **37**, 777 (1955).
280. ARDEN, T. V., BURSTALL, F. H., DAVIES, G. R., LEWIS, J. A. and LINSTEAD, R. P. *Nature* **162**, 691 (1948).
281. SLAMA, O. and LADENBAUER, I. M. *Mikrochim. Acta* 1238 (1956).
282. FIŞEL, S., GABE, I. and PONI, M. *Acad. Rep. Populare Romane, Fil. Iasi, Stud. Cercetari Stiint. Chim.* **13**, 151 (1962).
283. ADER, D. and PRIMOV, M. *Anal. Chim. Acta* **31**, 191 (1964).
284. CERRAI, E. and GHERSINI, G. *J. Chromatog.* **16**, 258 (1964).
285. MUNSHI, K. N. and DEY, A. K. *J. Prakt. Chem.* **18**, 233 (1962).
286. FIŞEL, S., GABE, I. and PONI, M. *Acad. Rep. Populare Romane Fil. Iasi Stud. Cercetari Stiint. Chim.* **13**, 33 (1962).
287. MAGEE, R. J. and SCOTT, I. A. P. *Talanta* **3**, 131 (1959).
288. LACOURT, A., SOMMEREYNS, G., DE GEYNDT, E., BARUH, J. and GILLARD, J. *Nature* **163**, 999 (1949).
289. LACOURT, A., SOMMEREYNS, G. and DE GEYNDT, E. *Mikrochem.* **36**, 312 (1951).
290. LACOURT, A., SOMMEREYNS, G., DE GEYNDT, E., BARUH, J. and GILLARD, J. *Mikrochem.* **34**, 215 (1949).
291. LACOURT, A., SOMMEREYNS, G., DE GEYNDT, E. and JACQUET, O. *Mikrochem.* **36**, 117 (1951).
292. LACOURT, A., SOMMEREYNS, G. and WANTIER, G. *Analyst* **77**, 943 (1952).
293. OLENOVICH, N. L. and WEI-I-CH'I *Zh. Analit. Khim.* **17**, 522 (1962).
294. MODREANU, F., FIŞEL, S. and CARPOV, A. *Acad. Rep. Populare Romane, Fil. Iasi, Stud. Cercetari Sciint.* **7**, 25 (1956).
295. ANDERSON, J. R. and LEDERER, M. *Anal. Chim. Acta* **4**, 513 (1950).
296. POLLARD, F. H., NICKLESS, G. and BANISTER, A. J. *Analyst* **81**, 577 (1956).
297. BIGHI, C. *Ann. Chim. (Roma)* **45**, 1087 (1955).
298. BUCHAR, E. and SUCHÝ, K. *Magyar Kém. Foly* **64**, 45 (1958).
299. BAYKUT, F. and BOLAYIR, S. *Istanbul Univ. Fen. Fak. Mec., Ser. C* **25**, Nos. 3-4 109 (1960).
300. IMAI, H. *J. Chem. Soc. Japan Pure Chem. Sect.* **80**, 761 (1959).
301. AGRINIER, H. *Compt. Rend.* **253**, 280 (1961).
302. OKÁČ, A. and ČERNÝ, P. *Chem. Listy* **46**, 14 (1952).
303. SIKORSKA-TOMICKA, H. *Chem. Analit. (Warsaw)*, **6**, 771 (1961).
304. HARTKAMP, H. and SPECKER, H. *Naturwiss.* **42**, 534 (1955).
305. LEDERER, M. *Anal. Chim. Acta* **8**, 259 (1953).
306. LACOURT, A. and SOMMEREYNS, G. *Mikrochim. Acta* 604 (1954).
307. STEVENS, H. M. *Anal. Chim. Acta* **15**, 51 (1956).
308. KRAUSZ, I. *Magyar Kém. Foly.* **66**, 296 (1960).
309. LACOURT, A. *Mikrochim. Acta* 824 (1955).
310. LEDERER, M., MOSCATELLI, V. and PADIGLIONE, C. *J. Chromatog.* **10**, 82 (1963).
311. SUCHÝ, K. *Chem. Listy* **48**, 1084 (1954).
312. DE SA, C. E. *Anais. Assoc. Brasil, Quím.* **16**, 53 (1957).
313. JOHNSON, O. H. and KRAUSE, H. H. *Anal. Chim. Acta* **11**, 128 (1954).
314. SHU-WEI PANG and SHU-CHUAN LIANG *Acta Chim. Sinica* **28**, 148 (1962).
315. ELBEIH, I. I. M. and ABOU-ELNAGA, M. A. *Anal. Chim. Acta* **19**, 123 (1958).
316. SARMA, B. *Trans. Bose Research. Inst., Calcutta* **18**, 105 (1949).
317. CERRAI, E. and TESTA, C. *J. Chromatog.* **5**, 442 (1961).
318. CARLESON, G. *Acta Chem. Scand.* **8**, 1673 (1954).
319. MALECZKI, E. *Veszprémi Vegyipari Egyetem Tudományos Ulésszakanak Elöadassi* No. 2, 19 (1957).
320. CLANET, F. *J. Chromatog.* **6**, 85 (1961).
321. HAYEK, E. and DALLA TORRE, H. *Mikrochim. Acta* 1078 (1963).
322. ALMÁSSY, G. and VIGVÁRI, M. *Magyar Kém. Foly* **61**, 109 (1955).
323. ARDEN, T. V., BURSTALL, F. H. and LINSTEAD, R. P. *J. Chem. Soc.* 311 (1949).
324. SEILER, H., SCHUSTER, M. and ERLENMEYER, H. *Helv. Chim. Acta* **37**, 1252 (1954).
325. LEDERER, M. and RALLO, F. *J. Chromatog.* **7**, 552 (1962).
326. PETERSON, H. T., Jr., *Anal. Chem.* **31**, 1279 (1959).

327. RAAEN, H. P. and THOMASON, P. F. *Anal. Chem.* **27**, 936 (1955).
328. RAO, V. K. M. *J. Sci. Ind. Research (India)* **19 B**, 320 (1960).
329. BARNABAS, J. *Naturwiss.* **42**, 153 (1955).
330. KELEMEN, I. B. and SZARVAS, P. *Acta Chim. Acad. Sci. Hung.* **27**, 261 (1961).
331. ELBEIH, I. I. M. and GABRA, G. G. *Chemist-Analyst* **52**, 77 (1963).
332. QUENTIN, K. E. *Z. Lebensmittel. Unters-Forsch.* **95**, 305 (1952).
333. BLASIUS, E. *Chromatographische Methoden in der analytischen und präparativen anorganischen Chemie*, p. 283 Enke-Verlag, Stuttgart (1958).
334. BAUMANN, I. H. *Naturwiss.* **43**, 300 (1956).
335. PFEIL, E., FRIEDRICH, A. and WACHSMANN, TH. *Z. Anal. Chem.* **158**, 429 (1957).
336. QURESHI, M. and AKHTAR, I. *Anal. Chem.* **34**, 1341 (1962).
337. BARBIERI, R., BUSULINI, L. and BELLUCO, N. *Ricerca Sci.* **29**, 2118 (1959).
338. HILL-COTTINGHAM, D. G. *J. Chromatog.* **8**, 261 (1962).
339. NAGAI, H. *J. Chem. Soc. Japan Pure Chem. Sect.* **78**, 1334 (1957).
340. NAGAI, H. *Kumamoto J. Sci., Ser. A* **3**, 181 (1957).
341. ĆELAP, M. B., JANJIĆ, T. J. and ŠPANOVIĆ, Ž. F. *Bull. soc. chim. (Belgrade)* **25/26**, 527 (1960).
342. LACOURT, A., GILLARD, J. and VAN DER WALLE, M. *Nature* **166**, 225 (1950).
343. VAECK, S. V. *Anal. Chim. Acta* **12**, 443 (1955).
344. SINGH, E. J. and DEY, A. K. *J. Chromatog.* **2**, 95 (1959).
345. SINGH, E. J. and DEY, A. K. *Z. Anal. Chem.* **165**, 179 (1959).
346. SINGH, E. J. and DEY, A. K. *Z. Anal. Chem.* **178**, 356 (1960).
347. SINGH, E. J. and DEY, A. K. *J. Indian. Chem. Soc.* **37**, 235 (1960).
348. LEDERER, M. *Anal. Chim. Acta* **13**, 350 (1955).
349. STEFANOVIĆ, G. and JANJIĆ, T. *Anal. Chim. Acta* **11**, 550 (1954).
350. NAGAI, H. *Kumamoto J. Sci., Ser. A* **2**, 304 (1955).
351. NAGAI, H. *Kumamoto J. Sci., Ser. A* **3**, 81 (1957).
352. NAGAI, H. *Kumamoto J. Sci., Ser. A* **3**, 171 (1957).
353. NAGAI, H. *J. Chem. Soc. Japan Pure Chem. Sect.* **82**, 349 (1961).
354. ASHIZAWA, T. *Japan Analyst* **10**, 683 (1961).
355. KIEŁCZEWSKI, W. and SUPINSKI, *J. Chem. Analit. (Warsaw)* **8**, 59 (1963).
356. LACOURT, A. and HEYNDRYCKX, P. *Mikrochim. Acta* 1389 (1956).
357. ELBEIH, I. I. M. *Chemist-Analyst* **45**, 75 (1956).
358. WEISS, D. *Z. Anal. Chem.* **185**, 273 (1962).
359. JURSÍK, F. *J. Chromatog.* **19**, 450 (1965).
360. FERNANDO, Q. *Anal. Chim. Acta* **12**, 432 (1955).
361. ZIEGLER, M. and WINKLER, H. *Z. Anal. Chem.* **166**, 241 (1959).
362. GRIMALDI, M. and LIBERTI, A. *J. Chromatog.* **15**, 510 (1964).
363. ERDEM, B. *Rev. Fac. Sci. Univ. Istanbul, Ser. C* **20**, 332 (1955).
364. WEISS, D. *Chem. Zvesti* **16**, 302 (1962).
365. ZIEGLER, M. *Z. Anal. Chem.* **174**, 323 (1960).
366. VAN ERKELENS, P. C. *Anal. Chim. Acta* **25**, 570 (1961).
367. LEWANDOWSKI, A., KIEŁCZEWSKI, W. and HOFFMANN, H. *Poznanskie Towarzystwo Przyjaciol. Nauk* **7**, 5 (1956).
368. ŠEDIVEC, V. and VAŠÁK, V. *Chem. Listy* **46**, 607 (1952).
369. IVANOV, S. A., MANEVA, D., TOLEVA, P. and MASHEV, M. P. *Chem. Abstr.* **54**, 24087 (1960).
370. NAGAI, H. *Chem. Abstr.* **11**, 869 (1964).
371. ĆELAP, M. B., JANJIĆ, T. J. and RADONOVIĆ, D. *Bull. soc. chim. (Belgrade)* **28**, 19 (1963).
372. MICHAL, J. and ZÝKA, J. *Chem. Listy* **51**, 56 (1957).
373. SHU-WEI-PANG, SHEN-CHANG-I and SHU-CHUAN-LIANG *Acta Chim. Sinica* **29**, 439 (1963).
374. LEDERER, M. *Nature* **163**, 598 (1949).
375. KEMBER, N. F. and WELLS, R. A. *Analyst* **76**, 579 (1951).
376. ANDERSON, J. R. A. and LEDERER, M. *Anal. Chim. Acta* **5**, 321 (1951).
377. NODDACK, W. and SANIR, S. *Chemiker Ztg.* **88**, 738 (1964).
378. RALLO, F. *J. Chromatog.* **8**, 132 (1962).
379. BAGLIANO, G. *J. Chromatog.* **17**, 168 (1965).
380. LEDERER, M. *Zh. Neorgan. Khim.* **3**, 1799 (1958).
381. SANIR, S. *Diss. Univ. Erlangen-Nürnberg* 1963.

382. VOISU, V. G. and NĂȘCUȚIU, T. *Acad. Rep. Populare Romane, Stud. Cercetari Chim.* **9**, 699 (1961).
383. KEMBER, N. F. and WELLS, R. A. *Analyst* **80**, 735 (1955).
384. BASOLO, F., LEDERER, M., OSSICINI, L. and STEPHEN, K. H. *Ric. Sci.* **32**, II-A, 485 (1962).
385. BASOLO, F., LEDERER, M., OSSICINI, L. and STEPHEN, K. H. *J. Chromatog.* **10**, 262 (1963).
386. DANON, J. and LEVI, M. C. *J. Chromatog.* **3**, 193 (1960).
387. LAUER, R. S. and POLUEKTOV, N. S. *Zavodsk. Lab.* **25**, 391 (1959).
388. LEDERER, M. *Compt. Rend.* **236**, 1557 (1953).
389. PANG-SHU-WEI and LIANG-SHU-CHUAN *Ko' Hsueh Tung Pao No.* **11**, 46 (1962); *Hua Hsueh Pao* **30**, 237 (1964).
390. CERRAI, E. and TRIULZI, C. *J. Chromatog.* **16**, 365 (1964).
391. CERRAI, E. and TESTA, C. *J. Chromatog.* **8**, 232 (1962).
392. POLLARD, F. H., McOMIE, J. F. W. and STEVENS, H. M. *J. Chem. Soc.* 3435 (1954).
393. LEDERER, M. *Nature* **176**, 462 (1955).
394. LEDERER, M. *Anal. Chim. Acta* **15**, 122 (1956).
395. LEDERER, M. *Anal. Chim. Acta* **15**, 46 (1956).
396. NAGAI, H. *Bull. Chem. Soc. Japan* **33**, 715 (1960).
397. ALBERTI, G. and MASSUCI, M. A. *J. Chromatog.* **11**, 394 (1963).
398. LAUER, R. S. and POLUEKTOV, N. S. *Zavodsk. Lab.* **25**, 391 (1959).
399. POLUEKTOV, N. S., LAUER, R. S. and YAGUYATINSKAYA, G. YA. *Redkozemelnye Elementy. Akad. Nauk SSSR* 199 (1958).
400. FIDLER, J. and MICHAL, J. *Rudy* **11**, Supplement No. 5, 35 (1963).
401. WEIL, H. *Can. Chem. Process. Ind.* **33**, 1036 (1949).
402. DE CARVALHO, R. A. G. *Rev. Port. Quím.* **1**, 137, 233, 355 (1958).
403. OSSICINI, L., SARACINO, F. and LEDERER, M. *J. Chromatog.* **16**, 524 (1964).
404. MATSUURA, M. and KOJIMA, M. *J. Chem. Soc. Japan Pure Chem. Sect.* **80**, 459. (1959).
405. MATSUURA, M. and KOJIMA, M. *J. Chem. Soc. Japan Pure Chem. Sect.* **80**, 1210 (1959).
406. KLINKEN, B. and LEDERER, M. *J. Chromatog.* **6**, 524 (1961).
407. ALLULLI, S. and GRASSINI, G. *J. Chromatog.* **7**, 414 (1962).
408. TAUBER, R. and SCHÖNFELD, T. *J. Chromatog.* **4**, 222 (1960).
409. LINDNER R. *Naturwiss.* **33**, 119 (1946).
410. LINDNER, R. *Z. Naturforsch.* **2a**, 329 (1947).
411. SAVIĆ, P. and CVJETIĆANIN, D. N. *Bull. Inst. Nuclear Sci. Boris. Kidrič (Belgrade)* **4**, 21 (1954).
412. CONTE, P., MUXART, R. and ARAPAKI, H. *J. Chromatog.* **3**, 96 (1960).
413. LEDERER, M. *J. Chromatog.* **1**, 172 (1958).
414. LEDERER, M. and VERNOIS, J. *Compt. Rend.* **244**, 2388 (1957).
415. JAKOVAC, Z. and LEDERER, M. *J. Chromatog.* **1**, 289 (1958).
416. JAKOVAC, Z. and LEDERER, M. *J. Chromatog.* **2**, 411 (1959).
417. CLANET, F. *J. Chromatog.* **7**, 373 (1962).
418. KELLER, C. *J. Chromatog.* **7**, 535 (1962).
419. COWAN, M. R. and FOREMAN, J. K. *Chem. Ind. London* 1583 (1954).
420. FINK, R. M. and FINK, K. F. *U. S. At. Energy Comm. Rep.* UCLA-30 (1949).
421. ESCHRICH, H. *Kjeller Rept.* KR 11, 3 (1961); *Chem. Abstr.* **57**, 101 (1962), C. 10, 1962.
422. MIKULSKI, J. and STRONSKI, I. *Nukleonika* **6**, 775 (1961).
423. AGRINIER, H. *Bull. Soc. Franc. Minéral Crist.* **80**, 181 (1957).
424. MAZZA, L. and SARDO, L. *Gazz. Chim. Ital.* **93**, 1155 (1963).
425. PIERCE, T. B. and PECK, F. P. *J. Chromatog.* **6**, 248 (1961).
426. KALJANIN, A. V. *Radiochimiya* **5**, 749 (1963).
427. LIANG-SHU-CHUAN *Acta Chim. Sinica* **24**, 205 (1958).
428. VOLMAR, Y., EBEL, J.-P. and YACOUB, F. B. *Compt. Rend.* **235**, 372 (1952).
429. THILO, E. and GRUNZE, H.; according to HETTLER, H. *J. Chromatog.* **1**, 389 (1958).
430. GASSNER, K. *Mikrochim. Acta* 594 (1957).
431. PFRENGLE, O. *Z. Anal. Chem.* **158**, 81 (1957).

432. ANDO, T., ITO, J., ISHII, S. and SODA, T. *J. Chem. Soc. Japan Pure Chem. Sect.* **25**, 78 (1952).
433. EBERT, M. and VARHANÍKOVÁ, A. *Chem. Průmysl* **12**, 192 (1962).
434. MILLER, C. C. and MAGEE, R. J. *J. Chem. Soc.* 3183 (1951).
435. POLLARD, F. H. and NICKLESS, G. et al. *J. Chromatog.* **9**, 485, 493, 506, 5, 13 (1962); **10**, 73, 78 (1963); **11**, 383, 534 (1963); **12**, 527 (1963); **16**, 207 (1964).
436. CVJETIĆANIN, N. M. and OBRENOVIĆ, I. D. *Bull. Inst. Nucl. Sci. Boris Kidrić Belgrade* **11**, 173 (1961).
437. KOLLOFF, R. H. *Anal. Chem.* **33**, 373 (1961).
438. GAUTHIER, P. *Bull. Soc. Chim. France*, 981 (1955).
439. EBEL, J. P. and VOLMAR, I. *Compt. Rend.* **233**, 415 (1951).
440. LEDERER, M. *Anal. Chim. Acta* **11**, 524 (1954).
441. EBEL, J. P. *Compt. Rend.* **234**, 732 (1952).
442. EBEL, J. P. *Bull. Soc. Chim. Biol.* **34**, 330 (1952).
443. EBEL, J. P. *Bull. Soc. Chim. Biol.* **34**, 321 (1952).
444. EBEL, J. P. *Compt. Rend.* **234**, 621 (1952).
445. EBEL, J. P. *Bull. Soc. Chim. France* 1089 (1953).
446. EBEL, J. P. *Bull. Soc. Chim. France* 998 (1953).
447. EBEL, J. P. *Bull. Soc. Chim. France* 991 (1953).
448. EBEL, J. P. *Bull. Soc. Chim. France* 1096 (1953).
449. EBEL, J. P. *Mikrochim. Acta* 679 (1954).
450. EBEL, J. P. COLAS, J. and BUSCH, N. *Bull. Soc. Chim. France* 1087 (1955).
451. SERVIGNE, Y. and DUVAL, C. *Compt. Rend.* **245**, 1803 (1957).
452. POLLARD, F. H., NICKLESS, G. and GLOVER, R. B. *J. Chromatog.* **15**, 518 (1964).
453. POLLARD, F. H., McOMIE, J. F. W. and JONES, D. J. *J. Chem. Soc.* 4337 (1955).
454. POLLARD, F. H., JONES, D. J. and NICKLESS, G. *J. Chromatog.* **15**, 393 (1964).
455. POLLARD, F. H., NICKLESS, G., JONES, D. J. and GLOVER, R. B. *J. Chromatog.* **15**, 407 (1964).
456. POLLARD, F. H., NICKLESS, G. and GLOVER, R. B. *J. Chromatog.* **15**, 533 (1964).
457. SCOFFONE, E. and CARINI, E. *Chem. Abstr.* **50**, 729 (1956).
458. BIGHI, C., TROBANELLI, G. and PANCALDI, G. *Bull. Sci. Fac. Chim. Ind. Bologna* **16**, 92 (1958).
459. COCH-FRUGONI, J. A. *J. Chromatog.* **1**, 90 (1958).
460. SERVIGNE, Y. *Compt. Rend.* **239**, 272 (1954).
461. SERVIGNE, Y. *Mikrochim. Acta* 750 (1956).
462. BROOMHEAD, J. A. and GIBSON, N. A. *Anal. Chim. Acta* **26**, 265 (1962).
463. STAMM, H. H. and SCHROEDER, H. J. *J. Chromatog.* **10**, 392 (1963).
464. LEDERER, M. *Science* **110**, 115 (1949).
465. BURSTALL, F. H., DAVIES, G. R., LINSTEAD, R. P. and WELLS, R. A. *J. Chem. Soc.* 516 (1950).
466. ANDO, T. and ISHII, S. *Bull. Chem. Soc. Japan* **25**, 106 (1952).
467. IVANOV, S. A. *Compt. Rend. Acad. Bulgare Sci.* **12**, 41 (1959).
468. YAMADA, S. *J. Chem. Soc. Japan Pure Chem. Sect.* **82**, 35 (1961).
469. HEYNCZAK, A. J., KNYPL, J. S. and ANTOSZEWSKI, R. *Chem. Analit. (Warsaw)* **5**, 407 (1960).
470. BORN, H. J. and STÄRK, H. *Atomkern Energie* **718**, 286 (1959).
471. NAEUMANN, R. *J. Chromatog.* **18**, 385 (1965).
472. ELBEIH, I. I. M. and GABRA, G. G. *Chemist-Analyst* **52**, 36 (1963).
473. DELOACH, W. S. and DRINKARD, C. *J. Chem. Educ.* **28**, 461 (1951).
474. PEREY, M. and ADLOFF, J. P. *Compt. rend.* **236**, 1163 (1953).
475. PEREY, M. and ADLOFF, J. P. *Compt. rend.* **239**, 1389 (1954).
476. RAO, C. L. and GUPTA, A. R. *J. Chromatog.* **5**, 147 (1961).
477. NEALES, T. F. *J. Chromatog.* **16**, 262 (1964).
478. MURATA, A. *J. Chem. Soc. Japan Pure Chem. Sect.* **77**, 781 (1956).
479. GRASSINI, G. and OSSICINI, L. *J. Chromatog.* **7**, 351 (1962).
480. OSSICINI, L. *J. Chromatog.* **9**, 114 (1962).
481. LEDERER, M. and OSSICINI, L. *J. Chromatog.* **15**, 514 (1964).
482. NAKANO, S. *J. Chem. Soc. Japan Pure Chem. Sect.* **74**, 56 (1953).
483. BISHOP, J. R. *Analyst* **81**, 291 (1956).
484. FOUARGE, J. *Anal. Chim. Acta* **18**, 225 (1958).
485. FOUARGE, J. *Anal. Chim. Acta* **12**, 342 (1955).
486. GHE, A. M. *Ann. Chim. (Roma)* **50**, 1321 (1960).

487. Rees-Evans, D. B., Ryan, W. and Wells, R. A. *Analyst* **83**, 356 (1958).
488. Ghe, A. M. and Fiorentini, R. *Ann. Chim.* (*Roma*) **45**, 400 (1955).
489. Qureshi, M. and Rawat, J. P. *Sepn. Sci.* **6**, 451 (1971).
490. Kember, N. F. *Analyst* **77**, 78 (1952).
491. Williams, A. F. *Analyst* **77**, 297 (1952).
492. Kingsbury, G. W. J. and Temple, R. B. F. *Analyst* **77**, 307 (1952).
493. Ryan, W. and Williams, A. F. *Analyst* **77**, 293 (1952).
494. Muzzarelli, R. A. A. and Bate, L. C. *Talanta* **12**, 823 (1965).
495. Chew, B. and Lindley, G. *Metallurgia* **53**, 45 (1956).
496. Korenman, I. M. and Krainova, Z. V. *Zh. Prikl. Khim.* **19**, 604 (1946).
497. Marjanović-Krajovan, V., Turina, S., Indjić, N. and Jovanović, V. Z. *Anal. Chem.* **195**, 427 (1963).
498. Winchester, J. W. *J. Chromatog.* **10**, 502 (1963).
499. Sochacka, R. J. and Siekierski, S. *J. Chromatog.* **16**, 376 (1964).
500. Siekerski, S. and Sochacka, R. J. *J. Chromatog.* **16**, 385 (1964).
501. Cheng, H. S., Ke, C.-H., Lin, C.-Y. and Winchester, J. W. *J. Chim. Chem. Soc.*, *Ser.* **2**, **10**, 80 (1963); *Z. Anal. Chem.* **207**, 437 (1965).
502. Pierce, T. B. and Peck, P. F. *Nature* **195**, 597 (1962); 194, 84 (1962).
503. Pierce, T. B., Peck, P. F. and Hobbs, R. S. *J. Chromatog.* **12**, 81 (1963).
504. Hui-Chang Ma, Cha-Ming-Ni and Shu-Chuan Liang *Scientia* (*Peking*) **1**, 64 (1964); *Anal. Abstr.* **12**, 2671 (1965).
505. Mikulski, J. and Stronski, J. *J. Chromatog.* **17**, 197 (1965).
506. Cerrai, E. and Testa, C. *J. Chromatog.* **6**, 443 (1961).
507. Cerrai, E. and Testa, C. *Energia Nucleare* **8**, 510 (1961).
508. Fidelis, I. and Siekierski, S. *J. Chromatog.* **17**, 542 (1965).
509. Fidelis, I., and Siekierski, S. *J. Chromatog.* **5**, 161 (1961).
510. Hamlin, A. G., Roberts, B. J., Loughlin, W. and Walker, S. G. *Anal. Chem.* **33**, 1547 (1961).
511. Beranová, M. and Novák, M. *Collection Czech. Chem. Commun.* **30**, 1073 (1965).
512. Cesarano, C., Pugnetti, G. and Testa, C. *J. Chromatog.* **19**, 589, 594 (1965).
513. Elbeih, I. I. M. and Gabra, G. G. *Chemist-Analyst* **52**, 36 (1963).
514. Laskowski, D. E. and McCrone, W. C. *Anal. Chem.* **23**, 1579 (1951).
515. Šteffek, M. *Chem. Listy* **55**, 1221 (1961).
516. Kertes, S. and Lederer, M. *Anal. Chim. Acta* **15**, 543 (1956).
517. Carleson, G. *Acta Chem. Scand.* **8**, 1673 (1954).
518. Alberti, G. and Grassini, G. *J. Chromatog.* **4**, 83 (1960).
519. Alberti, G., Dobici, F. and Grassini, G. *J. Chromatog.* **8**, 103 (1962).
520. Ch-Chung-Chang *Acta Chim. Sinica* **31**, 355 (1965).
521. Hejtmánek, M. and Hozmová ,E. *Mikrochim. Acta* 95 (1966).
522. Covello, M. and Ciampa, G. *J. Chromatog.* **20**, 201 (1965).
523. Saxena, G. C. and Singh, D. R. *Indian J. Chem.* **2**, 456 (1964).
524. Singh, D. R. and Saxena, G. C. *Indian J. Chem.* **2**, 251 (1964).
525. Qureshi, M. and Kham, M. A. *J. Indian. Chem. Soc.* **41**, 673 (1964).
526. Gupta, S. S. and Mukerjee, D. *Z. Anal. Chem.* **226**, 201 (1967).
527. Olenovich, N. L., Ufimtseva, S. N. and Rogashko, M. M. *Zh. Analit. Khim.* **20**, 368 (1965).
528. Kielczewski, W. and Schneider, M. *Chem. Analit.* (*Warsaw*) **9**, 693 (1964).
529. Fojtík, M. and Koprda, V. *Chem. Zvesti* **20**, 180 (1966).
530. Koprda, V. and Fojtík, M. *Chem. Zvesti* **20**, 676 (1966).
531. Galateanu, I. *J. Chromatog.* **19**, 208 (1965).
532. Bhatnagar, R. P., Siddique, S. and Rao, K. V. K. *Indian J. Chem.* **2**, 205 (1964).
533. Popper, E., Floreanu, J., Marcu, P. and Sosa, E. *Revue Roum. Chim.* **11**, 283 (1966).
534. Qureshi, M. and Akhtar, I. *Z. Anal. Chem.* **227**, 89 (1967).
535. Nagai, H. and Deguchi, T. *J. Chem. Soc. Japan, Pure Chem. Sect.* **84**, 499 (1963).
536. Ota, M. *Japan Analyst* **15**, 604 (1966).
537. Jursík, F. *J. Chromatog.* **26**, 339 (1967).
538. Kielczewski, W. and Uchman, W. *Chem. Analit.* (*Warsaw*) **10**, 915 (1965).
539. Năşcuţiu, T. and Grigorovici, A. *Revue Roum. Chim.* **10**, 449 (1965); *Studii Cercet. Chim.* **13**, 441 (1965).
540. Archundia, C. and Ajuria, S. *Revista Soc. Quím. Mex.* **10**, 77 (1966).

541. BHATNAGAR, R. P. and SHARMA, K. D. *Anal. Chim. Acta* **30**, 310 (1964).
542. CHING-HSIANG MAO *Acta Chim. Sinica* **31**, 432 (1965).
543. KEMPE, G. and DENN, W. *Z. Anal. Chem.* **217**, 169 (1966).
544. GUPTA, S. S. and MUKERJEE, D. *Z. Anal. Chem.* **213**, 38 (1965).
545. CHU-CHUN-CHANG and HSIU-HSIA YANG *Acta Chim. Sinica* **31**, 182 (1965).
546. QURESHI, M. and KHAN, F. *Anal. Chem.* **39**, 1329 (1967.)
547. DATTA, S. K. and SAHA, S. N. *Z. Anal. Chem.* **202**, 332 (1964).
548. NĂSCUŢIU, T. *Revue Roum. Chim.* **9**, 283 (1964); *Studii Cercet. Chim.* **12**, 283 (1964).
549. CHIA-LUNG KAO and SIN-HUA-CHANG *Acta Chim. Sinica* **29**, 421 (1963).
550. ELBEIH, I. I. M. and and ABOU-ELNAGA, M. A. *Chemist-Analyst* **55**, 43 (1966).
551. DAY, M. K. and CHAKRABARTTY, M. M. *Mikrochem. J.* **11**, 13 (1966).
552. CHIA-LUNG KAO and CHIU-YAO YUAN *Acta Chim. Sinica* **30**, 418 (1964).
553. SHU-WEI-PANG, CHIU-CHUAN LEI and SHU-CHUAN LIANG *Acta Chim. Sinica* **30**, 160 (1964).
554. ELISEEVA, G. D. and KULSAKAYA, O. A. *Ukr. Khim. Zh.* **32**, 220 (1966).
555. BAGLIANO, G. and LEDERER, M. *Ric., Scient. Riv.* **36**, 51 (1966).
556. RÖSSEL, T. and KIESSLICH, H. *Z. Anal. Chem.* **125**, 391 (1967).
557. GANCHEV, N. and KOEV, K. *Microchim. Acta* 97 (1964).
558. CHIH-HSIANG MAO *Acta Chim. Sinica* **30**, 496 (1964).
559. CHIH-HSIANG MAO *Acta Chim. Sinica* **30**, 412 (1964).
560. GANCHEV, N. and KOEV, K. *Mikrochim. Acta* 87 (1964).
561. HARRISON, B. L. and ROSENBLATT, D. H. *J. Chromatog.* **13**, 271 (1964).
562. GANCHEV, N. and KOEV, K. *Mikrochim. Acta* 92 (1964).
563. MALÝ, E. *J. Chromatog.* **19**, 206 (1965).
564. KIELCZEWSKI, W., SCHNEIDER, M. and TOMKOWIAK, J. *Chem. Analit.* (*Warsaw*), **11**, 343 (1966).
565. KIELCZEWSKI, W., SCHNEIDER, M. and UCHMAN, W. *Chem. Analit.* (*Warsaw*), **12**, 143 (1967).
566. KIELCZEWSKI, W. and SCHNEIDER, M. *Chem. Analit.* (*Warsaw*) **11**, 111 (1966).
567. KIELCZEWSKI, W. and UCHMAN, W. *Chem. Analit.* (*Warsaw*) **11**, 543 (1966).
568. BERTHELOT, C. A., HERRMANN, S., LAUER, K. F. and MUZZARELLI, R. A. A. *Euratom Rept.* EUR 578e (1964).
569. MUZZARELLI, R. A. A. *Talanta* **14**, 85 (1967).
570. MUZZARELLI, R. A. A. and MARCOTRIGIANO, G. *Talanta* **14**, 489 (1967).
571. MUZZARELLI, R. A. A. and BATE, L. C. *Talanta* **12**, 823 (1965).
572. FRITZ, J. S. and SCHMITT, D. H. *Talanta* **13**, 123 (1966).
573. MINCZEWSKI, J. and RÓZYCKI, C. *Chem. Analit.* (*Warsaw*) **10**, 965 (1965).
574. GROSSE-RUYKEN, H. and BOSKOLEN, J. *J. Prakt. Chem.* **30**, 77 (1965).
575. GROSSE-RUYKEN, H. and BOSKOLEN, J. *Kernenergie* **8**, 224 (1965).
576. ESCHRICH, H. *Z. Anal. Chem.* **226**, 100 (1967).
577. FRITZ, J. S. and LATWESEN, G. L. *Talanta* **14**, 529 (1967).
578. FRAZEE, R. T. *U. S. At. Energy Comm. Repr.* IS-T-2 (1965); *Chem. Abstr.* **63**, 1522h (1965).
579. FRITZ, J. S. and FRAZEE, R. T. *Anal. Chem.* **37**, 1358 (1965).
580. MOORE, F. L. and JURRIAANSE, A. *Anal. Chem.* **39**, 733 (1967).
581. O'LAUGHLIN, J. W., KAMIN, G. J., BERNER, D. L. and BANKS, C. V. *Anal. Chem.* **36**, 2110 (1964).
582. SMULEK, W. and SIEKIERSKI, S. *J. Chromatog.* **19**, 580 (1965).
583. SOCHACKA, R. J. and SIEKIERSKI, S. *J. Chromatog.* **16**, 376 (1964).
584. HUFF, E. A. *J. Chromatog.* **27**, 229 (1966).
585. GOLLER, E. J. *J. Chem. Educ.* **42**, 442 (1965).
586. PESCHKE, W. *J. Chromatog.* **20**, 572 (1965).
587. HASHMI, M. H. and ADIL, A. S. *Z. Anal. Chem.* **227**, 170 (1967).
588. GALATEANU, I. *J. Chromatog.* **19**, 208 (1965).
589. LESIGANG, M. and HECHT, F. *Mikrochim. Acta* 508 (1964).
590. SEILER, H. and SEILER, M. *Helv. Chim. Acta* **43**, 1939 (1960).
591. SEILER, H. and SEILER, M. *Helv. Chim. Acta* **44**, 939 (1961).
592. SEILER, H. and ROTHWERIER, W. *Helv. Chim. Acta* **44**, 941 (1961).
593. SEILER, H. and KAFFENBERGER, T. *Helv. Chim. Acta* **44**, 1282 (1961).
594. SEILER, H. *Helv. Chim. Acta* **44**, 1753 (1961).
595. SEILER, H. *Helv. Chim. Acta* **45**, 381 (1962).

596. SEILER, H. *Helv. Chim. Acta* **46**, 2629 (1963).
597. SEILER, H., BIERBRICHER, C. and ERLENMEYER, H. *Helv. Chim. Acta* **46**, 2636 (1963).
598. SEILER, H. and ERLENMEYER, H. *Helv. Chim. Acta* **47**, 264 (1964).
599. SEILER, H. and SEILER, M. *Helv. Chim. Acta* **48**, 117 (1965).
600. JANAUER, G. E. and JOHNSTON, R. C. *Anal. Chem.* **38**, 786 (1966).
601. GAGLIARDI, E. and LIKUSSAR, W. *Mikrochim. Acta* 765 (1965).
602. GAGLIARDI, E. and LIKUSSAR, W. *Mikrochim. Acta* 1053 (1965).
603. GAGLIARDI, E. and POKORNÝ, G. *Mikrochim. Acta* 577 (1966).
604. DANEELS, A., MASSART, D. L. and HOSTA, J. *J. Chromatog.* **18**, 144 (1965).
605. KAYASHI, K. and OGATA, T. *Japan Analyst* **14**, 1146 (1965).
606. PATRIARCHE, G. *J. Pharm. Belg.* **20**, 336 (1965).
607. YAMAOKA, K. and OKADA, K. *Japan Analyst* **15**, 67 (1966).
608. MENG-YEN WU and LIANG FANG WU *Scintia Sin.* 939 (1964).
609. SIECHOWSKI, J. *Chim. Analit. (Warsaw)* **9**, 391 (1964).
610. OGUMA, K. *Talanta* **14**, 685 (1967).
611. DRUDING, L. F. and HAGEL, R. B. *Anal. Chem.* **38**, 478 (1966).
612. BARK, L. S., GRAHAM, R. J. T. and McCORMICK, D. *Anal. Chim. Acta* **35**, 268 (1966).
613. PETROVIĆ, S. M. and CANIĆ, V. D. *Z. Anal. Chem.* **228**, 339 (1967).
614. CANIĆ, V. D., TURČIĆ, M. N., BUGARSKI-VOJINOVIĆ, M. B. and PERIŠIĆ, N. V. *Z. Anal. Chem.* **229**. 93 (1967).
615. TAKEUCHI, T. and TSUNODA, Y. *J. Chem. Soc. Japan, Pure Chem. Sect.* **87**, 251 (1966).
616. MUTO, M. *J. Chem. Soc. Japan, Pure Chem. Sect.* **85**, 782 (1964).
617. MUTO, M. *J. Chem. Soc. Japan, Pure Chem. Sect.* **86**, 91 (1965).
618. BANDLER, M. and STUHLMANN, F. *Naturwiss.* **51**, 57 (1964).
619. BANDLER, M. and MENGEL, M. *Z. Anal. Chem.* **211**, 42 (1965).
620. AURENGE, J., DEGEORGES, M. and NORMAND, J. *Bull. Soc. Chim. France* 508 (1964).
621. CANIĆ, V. D., TURČIĆ, M. N., PETROVIĆ, S. M. and PETROVIĆ, S. E. *Anal. Chem.* **37**, 1576 (1965).
622. RÖSSEL, T. and KREISSLICH, H. *Z. Anal. Chem.* **229**, 96 (1967).
623. DEGUCHI, T. *Japan Analyst.* **15**, 352 (1966).
624. HRANISAVLJEVIĆ-JAKOVLJEVIĆ, M. and PEJKOVIĆ-TADIĆ, I. *Mikrochim. Acta* 936 (1965).
625. PEJKOVIĆ-TADIĆ, I. and HRANISAVLJEVIĆ-JAKOVLJEVIĆ, M. *Mikrochim. Acta* 940 (1965).
626. SENF, H. J. *J. Chromatog.* **27**, 331 (1967).
627. SENF, H. J. *J. Chromatog.* **21**, 363 (1966).
628. CHINH-TE HU and CHENG-LI LIU *Acta Chim. Sinica* **31**, 267 (1965).
629. BARK, L. S., DUNCAN, G. and GRAHAM, R. J. *Analyst* **92**, 347 (1967).
630. LEDERER, M. and OSSICINI, L. *J. Chromatog.* **13**, 188 (1964).
631. OSSICINI L. and LEDERER, M. *J. Chromatog.* **17**, 387 (1965).
632. SHUKLA, S. K. *J. Chromatog.* **21**, 92 (1966).
633. SHU-WEI PANG and SHU-CHUAN LIANG *Acta Chim. Sinica* **30**, 401 (1964).
634. WAKSMUDSKI, A. and PRZESZLAKOWSKI, S. *Chem. Analit. (Warsaw)* **11**, 159 (1966).
635. KNOCH, W. and LAHR, H. *Radiochim. Acta* **4**, 114 (1965).
636. WERNER, G. *J. Chromatog.* **22**, 400 (1966).
637. CERRAI, E. and GHERSINI, G. *J. Chromatog.* **24**, 383 (1966).
638. O'LAUGHLIN, J. W., KAMIN, G. J. and BANKS, C. V. *J. Chromatog.* **21**, 460 (1966).
639. ZHI-TEI HU *Acta Chim. Sinica* **30**, 426 (1964).
640. O'LAUGHLIN, J. W., FERGUSON, J. W., RICHARD, J. J. and BANKS, C. V. *J. Chromatog.* **24**, 376 (1966).
641. ZHI-TEI HU and SIN-CHIUM SHI *Acta Chim. Sinica* **30**, 352 (1964).
642. QURESHI, M. and QURESHI, S. Z. *J. Chromatog.* **22**, 198 (1966).
643. RANDERATH, K. and RANDERATH, E. *J. Chromatog.* **22**, 110 (1966).
644. HOLZAPFEL, H., LE-VIET-LAN and WERNER, G. *J. Chromatog.* **20**, 580 (1966).
645. HOLZAPFEL, H., LE VIET-LAN and WERNER, G. *J. Chromatog.* **24**, 153 (1966).
646. ACKERMANN, G. and FREY, H.-P. *Z. Anal. Chem.* **233**, 321 (1968).
647. CVJETIĆANIN, N. *J. Chromatog.* **32**, 384 (1968).
648. JANARDHAN, P. D. and PAUL, A. *Sepn. Sci.* **2**, 597 (1967).

649. PAUL, A. and JANARDHAN, P. D. *Microchem. J.* **13**, 286 (1968).
650. CHIOTIS, E. L., WELFORD, G. A. and MORSE, R. S. *Mikrochim. Acta* 297 (1968).
651. QURESHI, M. and MATHUR, K. N. *Anal. Chim. Acta* **41**, 560 (1968).
652. QURESHI, M. and KHAN, F. *J. Chromatog.* **34**, 222 (1968).
653. TOUL, J. and TUSL, J. *Collection Czech. Chem. Commun.* **33**, 3703 (1968).
654. KIRIN, I. S. and ZASITOV, V. M. *Radiokhimiya* **10**, 42 (1968).
655. CRISCO, C. and LADA, H. F. *Rept. U. S. Army Nucl. Defense Lab.* AD-665357 (1968).
656. IONOVA, L. A. and POSTNIKOV, N. N. *Khim. Prom.* 198 (1969).
657. KIELCZEWSKI, W., GERA, J. and TOMKOWIAK, J. *Chim. Analit. (Warsaw)* **14**, 135 (1969).
658. KAWAMURA, S., KUROTAKI, K., KURAKU, H. and IZAWA, M. *J. Chromatog.* **26**, 557 (1967).
659. HASHMI, M. H., SHAHID, M. A. and CHUGHTAI, F. R. *Mikrochim. Acta* 309 (1968).
660. POKORNY, G. and BAYER, W. *Mikrochim. Acta* 582 (1968).
661. MASSART, D. L., SAINTE, G. and HOSTE, J. *Acta Chim. Hung.* **52**, 229 (1967).
662. OGUMA, K. *Talanta* **15**, 860 (1968).
663. VAGINA, N. S. and VOLYNETS, M. P. *Zh. Analit. Khim.* **23**, 521 (1968).
664. VAGINA, N. S. and FAKINA, L. K. *Zavodsk. Lab.* **34**, 928 (1968).
665. SENF, H. J. *Mikrochim. Acta* 954 (1968).
666. VUKCEVIC-KOVACEVIC, V. and SEREMET, S. *Sci. Pharm.* **36**, 26 (1968).
667. TRUJILLO, A. and FRYE, H. *Chemist-Analyst* **56**, 90 (1967).
668. HASHMI, M. H. and SUBHAN, A. A. *Mikrochim. Acta* 947 (1968).
669. OGUMA, K. *Talanta* **16**, 409 (1969).
670. DRĂGULESCU, C., FRUCHTER, S. and ZAHARIA, M. *Rev. Roum. Chim.* **12**, 139 (1967).
671. KURODA, K., KAWABUCHI, K. and ITO, T. *Talanta* **15**, 1486 (1968).
672. GAIBAKYAN, D. S. *Armyan. Khim. Zh.* **22**, 13 (1969).
673. LESIGANG-BUCHTELA, M. and BUCHTELA, K. *Mikrochim. Acta* 570 (1967).
674. VOLYNETS, M. P. and GUSEVA, L. I. *Zh. Analit. Khim.* **23**, 947 (1968).
675. SEILER, H. *Helv. Chim. Acta* **52**, 319 (1969).
676. ANDREEV, A. G., ZHEREKOV, V. G. and KUTSENKO, N. V. *Radiokhimiya* **11**, 80 (1969).
677. YAMABE, T., IIDA, T. and TAKAI, N. *Bull. Chem. Soc. Japan* **41**, 1959 (1968).
678. TANZER, J. M., KRICHEVSKY, M. I. and CHASSY, B. *J. Chromatog.* **38**, 526 (1968).
679. VAN OOIJ, W. J. and HOUTMAN, J. P. W. *Z. Anal. Chem.* **244**, 38 (1969).
680. TAKEUCHI, T. and TSUNODA, Y. *J. Chem. Soc. Japan, Pure Chem. Sect.* **88**, 176 (1967).
681. MUCHOVÁ, A. and JOKL, V. *Chem. Zvesti* **22**, 62 (1968).
682. TREHAN, J. C. *Chromatographia* **2**, 17 (1969).
683. VANDERDEELEN, J. *J. Chromatog.* **39**, 521 (1969).
684. WASILEWSKA, L. *Farm. Polska* **24**, 56 (1968).
685. HAWORTH, D. T. and ZETMEILS, M. *J. Sepn. Sci.* **3**, 145 (1968).
686. BABA, T., YONODA, H. and MUTO, M. *Bull. Chem. Soc. Japan* **41**, 1965 (1968).
687. SENF, H. J. *Mikrochim. Acta* 522 (1969).
688. TUSTANOWSKI, S. *J. Chromatog.* **31**, 270 (1967).
689. MUZZARELLI, R. A. A. and MARCOTRIGIANO, C. *Talanta* **14**, 489 (1967).
690. FRITZ, J. S. and LATWESEN, G. L., *Talanta* **14**, 529 (1967).
691. KALININ, S. K., KATYHIN, G. S., NIKITIN, M. K. and YAKOVLOVA, G. A. *Zh. Analit. Khim.* **23**, 1481 (1968).
692. HERRMANN, E. *J. Chromatog.* **38**, 498 (1968).
693. MOORE, F. L. *Anal. Chem.* **40**, 2130 (1968).
694. ZIV, D. M., SHESTAKOV, B. I. and SHESTAKOVA, I. A. *Radiokhimiya* **10**, 738 (1968),
695. FRITZ, J. S. and DAHMER, L. D. *Anal. Chem.* **40**, 20 (1968).
696. ESCHRICH, H. and HUNDERE, I. *Tech. Rept. Eurochemie Co.* No. ETR-222 (1968).
697. HEUNISCH, G. W. *Anal. Chim. Acta* **45**, 133 (1969).
698. THIELEMANN, H. *Mikrochim Acta* 524 (1971).
699. PILSON, M. E. Q. and FRAGALA, R. *J. Anal. Chim. Acta* **52**, 553 (1970).
700. THIELEMANN, H. *Mikrochim. Acta* 746 (1971).
701. SCHMID, E. R. *Mikrochim. Acta* 301 (1970).
702. FRITZ, J. S., BEUERMAN, D. R. and RICHARD, J. J. *Talanta* **18**, 1095 (1971).

703. MUZZARELLI, R. A. A. and SIPOS, L. *Talanta* **18**, 853 (1971).
704. MUZZARELLI, R. A. A. and TUBATINI, O. *Talanta* **16**, 1571 (1969).
705. FRITZ, J. S. and KENNEDY, D. C. *Talanta* **17**, 837 (1970).
706. KENNEDY, D. C. and FRITZ, J. S. *Talanta* **17**, 823 (1970).
707. FRITZ, J. S., FRAZEE, R. T. and LATWESEN, G. L. *Talanta* **17**, 857 (1970).
708. OVERMAN, R. F. *Anal. Chem.* **43**, 600 (1971).
709. FRITZ, J. S. and PETERS, M. A. *Talanta* **16**, 575 (1969).
710. FREI, R. W. and STOCKTON, C. A. *Mikrochim. Acta* 1196 (1969).
711. SCHMID, E. R. and PFANNHAUSER, W. *Mikrochim. Acta* 250 (1971).
712. CERRAI, E. and GHERSINI, G. *Analyst* **94**, 599 (1969).
713. HILGEMAN, F., SHIMOMURA, K. and WALTON, H. F. *Sepn. Sci.* **4**, 111 (1969)
714. ABE, S. and WEISZ, H. *Mikrochim. Acta* 550 (1970).
715. SHERMA, J. and VAN LENTEN, F. J. *Sepn. Sci.* **6**, 199 (1971).
716. GOURISSE, D. and CHESNÉ, A. *Anal. Chim. Acta* **45**, 311 (1969).
717. GOURISSE, D. and CHESNÉ, A. *Anal. Chim. Acta* **45**, 321 (1969).
718. LESIGANG-BUCHTELA, M. *Mikrochim. Acta* 1027 (1969).
719. FREI, R. and STILLMAN, H. *Mikrochim. Acta* 184 (1970).
720. GEIKE, Z. *Anal. Chem.* **258**, 284 (1972).
721. GAGLIARDI, E. and SHAMBRI, M. *Mikrochim. Acta* 154 (1970).
722. ŠOLJIĆ, Z., TURINA, S. and MARJANOVIĆ, V. *Z. Anal. Chem.* **258**, 231 (1972).
723. ŠOLJIĆ, Z., TURINA, S. and MARJANOVIĆ, V. *Mikrochim. Acta* 849 (1969).
724. JOHRI, K. N. and MEHRA, H. C. *Sepn. Sci.* **6**, 741 (171).
725. PHIPPS, M. *Anal. Chem.* **43**, 467 (1971).
726. JOHRI, K. N., KAUSHIK, N. K. and SINGH, K. *Mikrochim. Acta* 737 (1969).
727. HASHMI, M. H., CHUGHTAI, N. A., AHMAD, I. and AFZAL, A. U. *Mikrochim. Acta* 924 (1971).
728. HASHMI, M. H. and CHUGHTAI, N. A. *Mikrochim. Acta* 924 (1971).
729. VOLYNETS, M. P. and ERMAKOV, A. N. *Usp. Khim.* **39**, 934 (1970).
730. DUBUQUOY, C., GUSMINI, S., POUPARD, D. and VERRY, M. *J. Chromatog.* **57**, 455 (1971).
731. LEENE, H. R., DE VRIES, G. and BRINKMAN, U. A. TH. *J. Chromatog.* **57**, 173 (1971).
732. SHIMIZU, T. and ISHIKURA, R. *J. Chromatog.* **56**, 95 (1971).
733. BRINKMAN, U. A. TH. and DE VRIES, G. *J. Chromatog.* **56**, 102 (1971).
734. LIDA, T. and YAMABE, T. *J. Chromatog.* **56**, 373 (1971).
735. BRINKMAN, U. A. TH., STERRENBURG, P. J. J. and DE VRIES, G. *J. Chromatog.* **54**, 449 (1971).
736. RAAEN, H. P. *J. Chromatog.* **53**, 605 (1970).
737. OGUMA, K. and KURODA, R. *J. Chromatog.* **52**, 339 (1970).
738. ISHIDA, K. and KURODA, R. *Japan Analyst* **19**, 81 (1970).
739. KURODA, R., YOSHINUKI, N. and KAWABUCHI, K. *J. Chromatog.* **47**, 453 (1970).
740. SINIBALDI, M. and LEDERER, M. *J. Chromatog.* **51**, 556 (1970).
741. MASSOMI, Z. and HAWORTH, D. T. *J. Chromatog.* **48**, 581 (1970).
742. FREW, R. G. and PICKERING, W. F. *J. Chromatog.* **47**, 86 (1970).
743. QURESHI, M. and HUSAIN, W. *Sepn. Sci.* **4**, 197 (1969).
744. QURESHI, M. AKHTAR, I. and MATHUR, K. N. *Anal. Chem.* **39**, 1766 (1967).
745. QURESHI, M., MATHUR, K. N. and ISRAILI, A. H. *Talanta* **16**, 503 (1969).
746. LYLE, S. J. and NAIR, V. C. *Talanta* **16**, 813 (1969).
747. TESTA, C. *Anal. Chim. Acta* **50**, 447 (1970).

Chapter 7

CHROMATOGRAPHY
IN THE QUANTITATIVE ANALYSIS
OF INORGANIC COMPOUNDS

Chromatography is used not only in systematic or preliminary qualitative analysis, but also for preliminary separations in quantitative analysis. Methods of quantitative analysis based on the use of chromatography make it possible to analyse a wide variety of materials. Nevertheless, the application of chromatography is not fully developed, and cannot yet be used to determine all common ions. Chromatography has been used mainly for components difficult to determine by conventional analytical methods or for which it is hard to find a direct chemical method of determination. Thus, various methods of chromatographic determination of uranium, thorium and beryllium have been developed, but the applications of similar methods to the separation of copper, chromium or manganese etc. are not widely used, although they might be suitable for practical use. At the same time, the simplicity, speed and convenience of chromatography makes it suitable for control and test purposes.

In this book it is not intended to discuss every method in detail, but rather to select typical examples of chromatographic methods of determination of certain components or raw materials, metals and alloys, water and other inorganic materials.

7.1. Quantitative Paper Chromatography

7.1.1. Chromatographic Analysis of Minerals

Determination of Arsenic

For the determination of arsenic in natural arsenides, Agrinier [1] proposed the following procedure. After decomposition of the material with 10% sodium carbonate solution followed by nitric acid, the solution is chromatographed with a mixture of acetone, hexane, nitric acid and water (5 : 4 : 1 : 1). Arsenic(III) (R_F 0.25), after oxidation to arsenic(V) is detected with 3% potassium iodide solution, forming a brown spot. This spot is then evaluated semiquantitatively by comparison with a series of standard chromatograms. The method is suitable for the determination of arsenic in the range 0.1 – 1%.

Determination of Beryllium

Ader and Alon [3] proposed the following procedure for the determination of beryllium in pegmatites. The sample (20 – 100 mg) is weighed in a small platinum dish, extracted with 5 ml of hydrofluoric acid and evaporated to dryness on a water-bath. The evaporation is repeated with several 5-ml portions of the acid. The residue is then treated with 5 ml of $6M$ hydrochloric acid and the dish is left on a water-bath until the decomposition is complete. After cooling, the solution is transferred to a 25 ml volumetric flask and diluted to volume with water. (If the sample contains only traces of beryllium up to 1 g of sample is decomposed with an appropriate amount of hydrofluoric acid).

A strip of Whatman No. 1 chromatographic paper is impregnated by immersion in 5% EDTA solution, the excess of which is removed by blotting between two filter papers. The paper is then dried. Then 0.01 ml of sample solution (corresponding to 0.002 – 0.3 μg of Be) is applied with a micropipette on the start, the chromatogram is dried and then allowed to stand for 2 minutes in a tank containing a beaker of boiling water. The paper is then transferred to a chromatographic chamber and chromatographed with ethanol–$5M$ hydrochloric acid (9 : 1) or ethyl methyl ketone–water–hydrochloric acid (75 : 10 : 15). After development, the dried chromatogram is exposed to ammonia vapours and sprayed with 0.075% aqueous Eriochrome cyanine R solution. The spots at R_F 0.7 are evaluated by comparison with a series of standard chromatograms. The method is suitable for determinations in materials containing 0.005 – 6% of beryllium.

For the separation of beryllium Agrinier [4] recommends a mixture of acetone and nitric acid, and the following procedure. Finely ground material (1 g) is digested with 4 ml of hydrofluoric acid and 2 ml of sulphuric acid on a water-bath for 30 minutes, then evaporated to dryness. The residue is extracted with 4 ml of nitric acid, mixed with 0.5 ml of hydrogen peroxide, and allowed to stand for 15 minutes. It is then evaporated to dryness and the evaporation repeated several times, 4 ml of nitric acid being added between each evaporation. The residue is then extracted with 2.5 ml of 15% nitric acid and filtered after standing for 5 minutes, a suitable microfiltration device being used, and 0.025 ml of the clear filtrate is then chromatographed with a mixture of acetone and nitric acid (13 : 7). After drying, the spot is detected by means of a solution of 0.2 g of quinalizarin in 10 ml of pyridine and 90 ml of acetone. The beryllium content is determined by comparison with standard chromatograms.

Demetrescu and Cioaca [5] decomposed beryl and feldspars containing beryllium by fusion with sodium and potassium carbonates, extracted the melt with dilute hydrochloric acid, and removed silica by the usual treatment with hydrofluoric acid. The sample was then chromatographed with butanol saturated with $3M$ hydrochloric acid, and the spot of beryllium at R_F 0.30 was eluted from the paper with 1% sodium hydroxide solution. Beryllium was titrated with a solution of quinalizarin in acetone.

Determination of Tin

Tin can be determined in silicate material after separation from other ions with a mixture of acetone and 1-butanol containing hydrogen peroxide. The peroxide also ensures separation from niobium, molybdenum, tungsten and bismuth [13].

A 1-g sample containing $75-100$ µg of tin is decomposed with a mixture of hydrofluoric and nitric acid (3 : 2) and the reaction mixture is evaporated to dryness. The residue is extracted with 2 ml of dilute hydrofluoric acid (1 : 10), centrifuged, and the clear supernatant liquid transferred to a strip of chromatographic paper, Chromatography is carried out with a mixture of acetone, 1-butanol, water, hydrogen peroxide and hydrochloric acid (80 : 70 : 40 : 10 : 3). After separation the chromatogram is detected with 0.1% ethanolic solution of Gallion and a 0.1% solution of haematoxylin in 50% ethanol. The violet spot of tin at R_F 0.75 is compared with spots on a standard chromatogram.

Determination of Germanium

Ladenbauer and Slama [9] proposed the following procedure for the determination of germanium in zinc blende. Approximately 0.1 g of the sample is decomposed by fusing for 20 minutes in a nickel crucible with $1-2$ g of sodium peroxide. The melt is extracted with water and neutralized with sulphuric acid, with cooling in ice. The solution is made approximately $6N$ with respect to sulphuric acid, and the mixture is saturated for 45 minutes with hydrogen sulphide in the cold, and then allowed to stand overnight under an H_2S atmosphere. The precipitate is filtered off with a filter stick and washed with $6N$ sulphuric acid saturated with gaseous hydrogen sulphide. The filter-stick is introduced into a small porcelain crucible and boiled gently with ammonia. The solution is allowed to stand overnight, and is then transferred to a platinum crucible where it is evaporated to dryness. The residue is extracted with nitric acid and evaporated once more. Free nitric acid is eliminated by careful heating over a flame (the platinum crucible should not become glowing). The residue is extracted with ammonia, boilled and evaporated to small volume. The solution is transferred by means of a pipette to a 12-cm broad chromatographic paper strip. The crucible is repeatedly washed with boiling water and the solution is also applied to the chromatogram. Chromatography is carried out with a mixture of 1-butanol and 10% nitric acid. A sample of germanium is chromatographed simultaneously as standard. The control chromatogram is detected with phenylfluorone and the corresponding zone on the chromatogram of the analysed sample is cut out, eluted, and determined photometrically [10].

Determination of Copper, Cobalt and Nickel

Hunt and co-workers [11] proposed the following method for soil analysis. A finely ground sample (0.5 g) is fused with 1 g of potassium bisulphate for 1 minute. The cooled melt is dissolved in 2 ml of a mixture of acids

(50 ml hydrochloric and 5 ml of nitric acid, diluted with 100 ml of water) by careful heating on a water-bath for 10 minutes. After cooling, 0.01 ml of the clear solution is spotted on the paper. This is introduced into the vessel containing a beaker of boiling water. After three minutes the paper is transferred into a chromatographic chamber and chromatographed with a mixture of 15 ml of hydrochloric acid, 10 ml of water, and 75 ml of ethyl methyl ketone. The chromatogram is then placed for 2 minutes in a vessel saturated with ammonia vapour. The still wet chromatogram is detected by spraying from both sides with 0.1% rubeanic acid solution. After drying, the spots on the chromatogram are compared with standard chromatograms prepared from standard solutions with the addition of a suitable amount of potassium bisulphate. The method is suitable for geological prospecting.

Determination of Niobium

Hunt and co-workers [11, 11a] applied paper chromatography to geochemical prospecting. They proposed the following methods for the determination of niobium in soils. A 1-g sample in a polyethylene beaker is decomposed by evaporation with 5 ml of hydrofluoric acid on a water-bath. To the dry residue are added exactly 2 ml of dilute hydrofluoric acid (1 : 3). The solution is stirred, allowed to stand for 30 minutes. Then 0.05 ml of the solution is transferred to a chromatogram and allowed to dry. It is then chromatographed with a mixture of 85 ml of ethyl methyl ketone and 15 ml of hydrofluoric acid. After development, the chromatogram is allowed to dry and is exposed to ammonia vapour for 3 minutes. The chromatogram is then sprayed from both sides with tannin solution, dried, and niobium is estimated by comparison of the spot with spots on standard chromatograms prepared alongside the sample.

The decomposition may be also done by fusion of 1 g of sample in a platinum crucible with 4 g of potassium bisulphate for 1 – 2 minutes. The cooled melt is extracted with 5 ml of hydrofluoric acid and evaporated to dryness on a water-bath. The residue is mixed with exactly 2 ml of dilute hydrofluoric acid (1 : 3) and allowed to stand for 30 minutes. Then 0.02 ml is applied on the paper, which is then chromatographed as described above.

Determination of Lead [11]

A 1-g sample of finely ground soil is digested with 1 ml of dilute nitric acid (1 : 3) on a water-bath for one hour. After cooling, 0.01 ml of the clear solution is applied to chromatographic paper and allowed to dry in air for 30 minutes. Chromatographic development is carried out with a mixture of 19 ml of redistilled methanol and 1 ml of hydrochloric acid. After development, the chromatogram is allowed to dry in air for 30 minutes and is first exposed to ammonia vapour for 2 minutes, and then allowed to stand in air for 15 minutes. Detection is done with dithizone and the spot evaluated by comparison with standard chromatograms. The method is suitable for the determination of lead in geochemical prospecting.

Determination of Selenium

Weatherley [12] proposed a method suitable for the determination of selenium in silicate material, in electrolytic slime, and in industrial waste.

When silicates are analysed, samples weighing 0.5 g are fused with 1 g of anhydrous sodium carbonate in a small platinum crucible. Then 5 ml of dilute hydrochloric acid (5 : 3) are added to the cooled melt, and the mixture is digested on a water-bath for 30 minutes. After cooling, 0.05 ml is transferred to a strip of chromatographic paper and chromatographed in a mixture of ethyl methyl ketone and hydrofluoric acid (80 : 20). The developed chromatogram is dried, detected with thiourea, and the zone of selenium near the solvent front is compared with standard chromatograms.

When sediments or slimes from electrolysis are analysed, 0.5 g of sample is digested with 5 ml of 20% sodium carbonate solution for 30 minutes and 0.05 ml of the clear solution is then transferred to a chromatographic paper and chromatographed as above (analysis of silicates).

In the case of wastes the decomposition is chosen according to whether metallic selenium is present or not. If it is not, the procedure is the same as for electrolytic sediments. If metallic selenium is present — which is probable in the case of ash — 0.5 g of the sample is digested on a water-bath with 5 ml of 40% sodium hydroxide solution for 30 minutes. From the clear solution obtained, 0.05 ml is transferred to a strip of chromatographic paper and the further procedure is as above.

Gel'man [64] recommends the following method for the determination of selenium and tellurium in mineral matter. A 0.5 − 2 g sample is dissolved in nitric acid and selenium and tellurium are separated from accompanying elements by co-precipitation on ferric hydroxide or aluminium hydroxide. The precipitate is dissolved in nitric acid and selenium and tellurium are determined by paper chromatography, the solvent being butanol saturated with water containing 4% of ammonium nitrate and 3 − 3.5% of nitric acid. The spots are located with a 5% solution of stannous chloride in 5% hydrochloric acid and the amounts of selenium and tellurium are determined by photometry of the coloured zones in reflected light or by comparison with a series of standard chromatograms. Concentrations of the order of 1 ppm of each element can be determined in ores.

Determination of Tantalum [11]

One gram of a finely ground soil sample is mixed and evaporated to dryness with 5 ml of hydrofluoric acid in a polythene beaker. The residue is digested with 2 ml of dilute hydrofluoric acid (1 : 3) for 30 minutes, with stirring. From the clear supernatant liquid, 0.05 ml is transferred to a chromatographic paper and allowed to dry in air for at least one hour. If niobium is present the drying should be extended to 2 hours. Chromatography is then carried out with a mixture of hydrofluoric acid (2 ml), water (8 ml) and ethyl methyl ketone (90 ml) in a chromatographic chamber previously saturated for 30 minutes with the vapours of the solvent system. The developed chromatogram is removed from the tank and sprayed immediately with quinalizarin solution. It is then exposed to acetic acid vapours for several minutes and the zone of tantalum is evaluated by comparison with standard chromatograms. The method is suitable for geochemical prospecting purposes.

Viktorova and Saltykova [65] recommend a method which involves conversion of tantalum into a fluoride complex by evaporation of the sample solution in the presence of hydrofluoric acid, followed by paper chromatography with acetone—40% hydrofluoric acid—water (45 : 4 : 1) as solvent. After location of the tantalum spots with Crystal Violet solution, the amount of tantalum present is determined by visual comparison of the separated spots with those of standards. There is no interference from rare-earth metals or from niobium or 30 other common elements, in the determination in rocks and minerals, but tungsten interferes.

Determination of Thorium

The separation of thorium from uranium, rare-earth metals and other elements is a problem tackled by many workers. The applications of these methods to the analysis of minerals have also been investigated. Thus, for example, Purushottam [14] proposed a method of determination of thorium in minerals, soils, and low-grade ores, based on a separation with a mixture of acetone, hydrochloric acid and 2-thenoyltrifluoroacetone. The decomposition or preliminary treatment of the sample is dependent on the character of the material analysed. Allanite is decomposed by digestion in concentrated hydrochloric acid, on a water-bath. The digestion is repeated until the decomposition is complete. The sample is then evaporated to dryness on a water-bath, the residue is extracted with 2 ml of concentrated hydrochloric acid and 0.01 ml of the clear solution is taken for chromatographic separation.

Monazite and samarskite are decomposed by heating with concentrated sulphuric acid, on a sand-bath. The reaction mixture after the decomposition is extracted with ice-water and filtered. The pH of the filtrate is adjusted to 3, and thorium is precipitated together with rare earths in the form of the oxalates by the addition of oxalic acid. The oxalates are ignited to the oxides, which are then dissolved in concentrated hydrochloric acid and evaporated to dryness on a water-bath. The residue is extracted with 2 ml of concentrated hydrochloric acid and an aliquot is chromatographed.

Soils are decomposed by heating with hydrochloric acid and nitric acid on a water-bath and evaporating to dryness several times, with addition of more acid between evaporations. When the decomposition is finished, the residue is extracted with 2 ml of acid and an aliquot is used for chromatography. Ores are decomposed in a similar manner, except that hydrofluoric acid is also added to make the decomposition complete.

Chromatography is carried out with a mixture of 12 ml of methanol, 6 ml of acetone, and 2 ml of dilute hydrochloric acid (1 : 3) containing 0.04 g of 2-thenoyltrifluoroacetone. Detection is carried out with an alcoholic solution of alizarin at pH 2.4—2.8. The violet-blue zone of thorium behind the solvent front is evaluated by direct comparison with standard chromatograms.

Elbeih and Abou-Elnaga [15] recommended a mixture of dioxan, nitric acid and antipyrine for the chromatographic separation of thorium in monazites and determination by titration with EDTA. A 10-g sample of finely ground monazite sand is decomposed by heating with 25 ml of sulphuric acid on a sand-bath, with occasional stirring. After evaporation of the mixture to a syrup, the sample is carefully added to 100 ml of distilled water and cooled. The solution is filtered and the residue is digested again with 10 ml of acid. The combined filtrates are made weakly alkaline by the addition of a small amount of sodium hydroxide with constant stirring, and the white precipitate of hydrated oxides is washed several times with distilled water by decantation. After collection on a sintered-glass filter (porosity 4) the hydrated oxides are dissolved in 30 ml of hot 50% nitric acid and the filter is thoroughly washed, first with 10 ml and five times with 5 ml of hot acid. The filtrate is cooled, transferred to a 100-ml volumetric flask, and made up to volume with 50% nitric acid.

An aliquot is applied to the chromatographic paper and chromatographed with a mixture of 100 ml of dioxan, 1 ml of nitric acid, 1 g of antipyrine and 2.5 ml of distilled water. Standards are chromatographed on the same sheet of paper. After development, the spots are detected by spraying with a mixture of kojic acid and 8-hydroxyquinoline and exposing to ammonia vapour and are viewed under ultraviolet light. Thorium appears at R_F 0.78 as a yellow-green fluorescent spot, uranium at 0.81 as a black spot. As their separation is not perfect both spots are treated together. The combined spots are cut out and placed in a titration flask. Water (25 ml) is added, followed by 5 drops of $0.1M$ nitric acid and exactly 4 ml of $0.001M$ EDTA. The mixture is boiled gently for 5 minutes, cooled, and mixed with an accurately measured $5-6$ ml of $0.001M$ bismuth nitrate and 2 or 3 drops of 0.1% Catechol Violet solution as indicator. If the solution is violet, the pH is too low and the solution must be buffered to pH $2-3$ with dilute ammonia, to produce a clear blue solution. It is then titrated with $0.001M$ EDTA till clear yellow. A standard is also titrated.

Determination of Titanium

For the determination of titanium in ilmenite [7] the sample is decomposed by fusing and dissolving in the usual manner [8]. Titanium is precipitated as hydroxide and the precipitate is dissolved in *aqua regia* and the solution diluted and chromatographed; the solvent used is 0.05 g of EDTA dissolved in a mixture of 50 ml of $1M$ nitric acid, 50 ml of 1-butanol and 1 ml of acetylacetone. The titanium spot is detected with a mixture of kojic acid and 8-hydroxyquinoline, and the titanium is determined photometrically.

Determination of Uranium

The determination of uranium is probably the most widely studied application of paper chromatography in the analysis of minerals. Elbeih and Abou-Elnaga [15] determine uranium and thorium in monazites simultaneously. After decomposition, and chromatographic separation in a mixture of dioxan, nitric acid and antipyrine (see above), and detection of the control chromatogram with potassium ferrocyanide, the zone with the supposed uranium spot is cut out and extracted in a Soxhlet apparatus with ethanol, acidified with nitric acid, for at least 10 hours. The extract is evaporated to a volume of 150 ml, cooled, and mixed with 150 ml of 3% sodium carbonate solution. It is then evaporated to a volume of approx. 50 ml and cooled. To both the solution of the sample and of the standard which is run simultaneously, 2 g of sodium carbonate are added and allowed to dissolve at room temperature. Hydrogen peroxide (10 ml) is then added, producing a yellow colour due to the peroxo-complex of uranium. In the presence of thorium a gelatinous white precipitate is also formed. The solution is filtered (sintered glass crucible) and the crucible washed several times with 10% sodium carbonate solution, then the filtrate is diluted to standard volume with water. The amount of the uranium present is determined by measurement of the absorbance of the solution against a blank.

Hunt [11] recommends the following procedure for geochemical prospecting. A 1-g sample of finely ground soil is decomposed in a platinum crucible on a water-bath with 2 ml of nitric acid and 2 ml of hydrofluoric acid. It is then evaporated to dryness, 1 ml of dilute nitric acid (1 : 3), is added, and 0.05 ml of the clear solution is applied to the start of a paper strip and chromatographed with a mixture of 2 ml of nitric acid, 1 ml of water and 17 ml of ethyl acetate. After development the chromatogram is removed from the tank, allowed to dry in air, and detected by lightly spraying both sides with 5% aqueous solution of potassium ferrocyanide. The amount of the uranium is determined by comparison with calibration chromatograms.

Plamondon [66] determined uranium in geochemical samples by the following method. The sample is weighed into a porcelain crucible and heated for one hour at 500 °C to destroy organic matter and transferred to a 30 ml beaker. The sample is then decomposed by heating gently for one hour with 2 ml of a solution of 60 g of aluminium nitrate in 100 ml of nitric acid (1 + 3) and then evaporated to dryness. The residue is treated with 2 ml of nitric acid (1 + 3), and the supernatant solution is transferred to a small test-tube and centrifuged. A 0.1-ml aliquot of the sample solution is transferred to a strip of chromatographic paper and developed with a mixture of isobutyl ketone (75 ml) with concentrated nitric acid (2.5 ml), mixed 30 minutes before use. Uranium is then detected with a 0.1% solution of 1-(2-pyridylazo)-2-naphthol in ammoniacal methanol, The violet spots are compared with standards after 15 minutes. In the range 1-12 ppm of uranium, the standard deviation is 1.9 ppm.

Determination of Bismuth

For quantitative determination of bismuth in inorganic material two methods have been proposed. According to the first [5] the sample is decomposed with *aqua regia* and after repeated evaporation with hydrochloric acid silica is eliminated in the usual manner. The residue is fused with sodium and potassium carbonates and bismuth is precipitated from the final solution together with the elements of the second analytical group by hydrogen sulphide at pH 0.5. The sulphides are filtered off, washed with 2% potassium nitrate solution, then dissolved in nitric acid, and the solution is made up to 10 ml in a volumetric flask. This solution is chromatographed in 1-butanol saturated with $3M$ hydrochloric acid, and the spot of bismuth at R_F 0.61 is extracted from the paper and determined photometrically with thiourea.

The second method [6] recommends for the separation a mixture of 88 ml of acetone, 12 ml of water and 0.5 ml of hydrochloric acid (or 3 ml of hydrofluoric acid, if much iron is present). The spot of bismuth at R_F 0.90 is detected with 2,5-dimercapto-1,3,4-thiazole and evaluated by comparison with a series of standard chromatograms.

Determination of Gold [2]

For the analysis of ores the procedure is as follows. One gram of the sample is heated to glowing in a quartz crucible at 1000° for 15 minutes. *Aqua regia* (4 ml) is added and evaporated to dryness. The residue is mixed with 3 ml of nitric acid and evaporated once more to dryness. The residue is extracted with 2 ml of 10% nitric acid and after centrifuging submitted to chromatographic analysis.

The decomposition of laterite ores is simpler. One gram of the sample is decomposed with 3 ml of concentrated nitric acid and 0.2 ml of concentrated hydrochloric acid. After evaporation to dryness, 3 ml of concentrated nitric acid are added and the evaporation is repeated. The residue is extracted with 2 ml of 10% nitric acid and centrifuged to give a solution suitable for further analysis.

The solvent mixture for the chromatographic separation is composed of 30 ml of ethanol, 40 ml of ethyl acetate and 30 ml of water containing 1 g of antipyrine and 1 ml of nitric acid. Gold moves with the solvent front and when the separation is finished it is detected with a saturated solution of *p*-dimethylaminobenzylidenerhodanine in a mixture of ethanol and acetone (1 : 1). For semiquantitative determination, a visual comparison with standard chromatograms is sufficient. Neither silver nor mercury interferes, and platinum metals (R_F 0.75) are also well separated.

7.1.2. Analysis of Metals and Alloys

Analysis of Steels

Paper chromatography is suitable in certain steel analyses. Ackermann and Schlosser [16] used it as a fast industrial method for the classification of steels, as well as for qualitative or semiquantitative evaluation.

A suitable amount of steel is dissolved in *aqua regia*. The solution is evaporated to dryness and the residue extracted with hydrochloric acid, care being taken that the final acid concentration is approx. 10%. Two strips of chromatographic paper are spotted with 5 µl of the solution on each and chromatographed upwards with a mixture of hydrochloric acid, 1-butanol and ethyl acetate (25 : 30 : 45). The chromatograms are dried until all the hydrochloric acid is removed. The first chromatogram is sprayed with $2M$ sodium hydroxide and exposed to bromine vapour. It is then dried and sprayed with a solution of 1 g of rubeanic acid and 0.5 g of salicylaldoxime in 100 ml of ethanol, and eventually exposed to ammonia vapour. If much nickel and negligible amounts of chromium are present, recognition of the yellow colour of chromium is difficult. Therefore, it is advantageous to spray the chromatogram with a 10% hydrogen peroxide solution, whereupon the red spot of nickel disappears and chromium can be easily recognized.

The second chromatogram is cut approximately 1 cm behind the solvent front. The upper part is detected with a solution of 2.6 g of stannous chloride and 5.0 g of sodium thiocyanate in 100 ml of dilute hydrochloric acid (1 : 1) and the lower part is sprayed with a solution of Alizarin S (0.1 g) and chromotropic acid (1 g) in 100 ml of 50% ethanol and then exposed to ammonia vapour. The spots of aluminium (R_F 0.0), chromium (0.07), nickel (0.07), vanadium (0.20), manganese (0.25), titanium (0.35), copper (0.55), cobalt (0.60), and molybdenum (0.85) can be evaluated semiquantitatively by comparison with standard spots on control chromatograms. Venturello and Ghe [17] also used paper chromatography for steel analysis. The method proposed by these authors can be used for the determination of nickel, copper, manganese, molybdenum, vanadium, chromium, titanium and cobalt in amounts up to 0.1%, using 1-butanol—acetoacetic acid—hydrochloric acid as solvent.

For the determination of niobium and tantalum in steel Scott and Magee [18] proposed the following method. One gram of steel is dissolved in 100 ml of a mixture of equal parts of 37% hydrochloric and 70% nitric acid. The solution is mixed with 100 ml of 71% perchloric acid and refluxed for 30 minutes. After cooling, 100 ml of water are added followed by 100 ml of saturated aqueous sulphur dioxide solution. The mixture is refluxed for 2 hours, cooled and filtered, and the residue on the filter is washed with 2% hydrochloric acid. The filter is ignited for 1 hour. The oxides formed are transformed into fluorides by evaporation with hydrofluoric acid mixed with a few drops of concentrated nitric acid. The residue is extracted with 1 ml of hydrofluoric acid and 0.05 ml is applied to a chromatographic paper and chromatographed with a mixture of methyl isobutyl ketone and 2% hydrofluoric acid (20 : 1). Niobium and tantalum are detected with $5M$ ammonia, followed by 5% 8-hydroxyquinoline solution in a mixture of methanol, chloroform and water (48 : 48 : 4). The chromatogram is dried in an oven at 120° for 30 minutes and the excess of 8-hydroxyquinoline is removed by washing the chromatogram with water. After further drying in the oven for 30 minutes the spots of niobium (R_F 0.23) and tantalum (R_F 0.72) are cut out from the paper and extracted by shaking the paper sections with 10 ml of hot $2M$ hydrochloric acid in separating-funnels. Chloroform is used to extract the niobium oxinate from the first funnel after adjustment of the pH to 8.5, with ammonia (indicator paper). The niobium is determined by measurement of the absorbance of the chloroform solution at 385 nm. Chloroform at pH 10 is used for the extraction of tantalum oxinate from the second funnel. The adjustment of pH and the extraction are done as described for niobium. The absorbance is measured at 390 nm.

Analysis of Cobalt Alloys

Dias and Coelho [19] used paper chromatography for the determination of metallic impurities in cobalt. On a Whatman No. 3 MM paper washed with a mixture of acetone, water and hydrochloric acid (20 : 20 : 1) and finally with water, 0.05−0.10 ml of a cobalt solution in $6M$ hydrochloric acid is applied and developed by the ascending method with a mixture of 50 ml of 1-butanol saturated with $3M$ hydrochloric acid, 10 ml of 2-propanol, 15 ml of acetone and 30 ml of benzyl alcohol, or with a mixture of 70 ml of 1-butanol and 30 ml of hydrochloric acid. With the first solvent the sequence of the separated ions is Zn (R_F 0.69), Fe (0.68), Pb (0.29), Cu (0.16) and Ni and Co (R_F 0.08). In the second the sequence is Fe (0.92), Zn (0.79), Cu (0.65), Pb (0.52), Co (0.51) and Ni (0.09). After the detection of zinc with 0.5% 8-hydroxyquinoline solution in 60% ethanol, iron with 10% potassium ferrocyanide solution, copper with 0.2% rubeanic acid solution, nickel and cobalt with rubeanic acid after the addition of ammonia, and lead with a 1% potassium thiocyanate solution in ethanol, the spots can be evaluated semiquantitatively by comparison with standard chromatograms.

Vaeck [20] proposed a direct densitometric determination of cobalt in alloys at a minimum concentration of 0.2%. The chromatogram is run in a mixture of acetone, 1-butanol, hydrochloric acid and acetylacetone (56 : 30 : 12 : 2). The separated cobalt is between copper (higher R_F value) and nickel. It is detected with rubeanic acid and determined by direct photometry on the chromatogram.

For the determination of traces of iron in tungsten-copper alloys a procedure also based on the use of paper chromatography has been proposed [21]. A weighed sample is decomposed by fusing with sodium carbonate, and tungsten is removed by evaporation with an acid. If the ratio of iron to cobalt is lower than 1 : 10 the iron can be determined by direct photometry with thiocyanate with an error of less than 0.3%. The best conditions for separation are obtained by using a mixture of acetone, hydrochloric acid and water (87 : 4 : 9). The iron is then eluted from the paper with dilute hydrochloric acid and determined photometrically with thiocyanate. The relative error for an Fe : Co ratio of 1 : 50−1 : 3000 is 0.03−3%.

Analysis of Aluminium Alloys

For the analysis of aluminium alloys, Venturello and Ghe [22] proposed circular paper chromatography. Copper, iron, silicon, magnesium, manganese, zinc, lead and nickel are determined as follows.

A 0.5-g sample of the alloy is dissolved in 7−8 ml of $6M$ hydrochloric acid in the cold, followed by gentle heating to complete dissolution. After making up to volume, an aliquot is taken for chromatography with a mixture of 1-butanol and $2M$ hydrochloric acid (1 : 1). The circular chromatogram is cut into four parts and these are detected with 1.2% dimethylglyoxime solution in ethanol (for the determination of nickel and iron), 1.2% ethanolic α-benzoinoxime solution (for copper and iron), 0.08% dithizone solution in chloroform (for zinc, lead and manganese), and a solution of quinalizarin (for magnesium). The width of the zones is proportional to the quantity of element.

Ackermann and Schlosser [63] analysed aluminium alloys by circular chromatography on papers 20 mm in diameter. After chromatography in ethyl acetate−1-butanol−hydrochloric acid (9 : 6 : 5) the paper was cut into four parts and detection carried out with the following reagents (elements and R_F values in brackets): dithio-oxamide (Ni 0.2, Cu 0.7); $2M$ sodium hydroxide and exposure to bromine vapour (Co 0.2, Mn 0.4); the sodium salt of chromotropic acid (Ti 0.46) — the solvent front of this part is also treated with potassium ferrocyanide (Fe 1.0); 0.01% dithizone in chloroform, followed by exposure to ammonia vapour (Zn 0.95). By using this method the authors determined the components of aluminium alloys in amounts down to 0.05%.

A determination of copper in aluminium and aluminium bronze was developed by Munshi and Dey [23]. The aluminium alloy is dissolved in concentrated nitric acid and the elements present are transformed into sulphates by evaporation with sulphuric acid. An aliquot of the solution is applied to the chromatographic paper and developed in a mixture of 75 ml of 1-butanol, 15 ml of hydrochloric acid, and 10 ml of water. The chromatogram is dried and a control chromatogram is detected with Chromazurol S to locate the copper spots. These are then eluted from the paper by boiling with water acidified with hydrochloric acid. The elution is repeated three times. To the combined extracts are added 20 ml of $5 \times 10^{-4} M$ Chromazurol S, and the pH is adjusted to 6.4. After 15 minutes the absorbance is measured at $520 - 600$ nm.

The separated aluminium can also be determined quantitatively if necessary. After elution from the paper (as for copper) and addition of 20 ml of $2.5 \times 10^{-4} M$ Chromazurol S (pH adjustment to 6.2 and dilution to 50 ml), the absorbance is measured at $520 - 580$ nm.

Analysis of High-Purity Metals

Šteffek [24] developed a method for the determination of trace metals (0.001%) in high-purity metals [24a — 24d]. For the determination of iron and copper in high-purity gold or antimony, the sample is dissolved in nitric acid and chromatographed with a mixture of 1-propanol, ethyl acetate, water and nitric acid (30 : 40 : 5 : 15) as solvent. A small amount of hydrochloric acid added to the solvent improves the separation. A control chromatogram is detected with a 2% ethanolic solution of salicylic acid and the corresponding zones are cut out, eluted with a mixture of nitric and hydrochloric acids and determined photometrically or polarographically. The determination of lead in high-purity bismuth is similar. Here the solvent is acetone, water and nitric acid (25 : 10 : 10) and the control chromatogram is detected with 1% aqueous thiourea solution.

When cobalt and zinc are determined in high-purity nickel, or when nickel is determined in high-purity cobalt, or cobalt and nickel in high-purity manganese, the metallic sample is first freed from mechanical impurities by washing with concentrated hydrochloric acid, followed by distilled water. The sample is then dissolved in hot 50% nitric acid (with boiling for nickel), the solution is evaporated until the nitrates crystallize, and the residue is diluted to a concentration of 1 g of sample per $7 - 8$ ml of solution. For the determination of zinc and cobalt in nickel the solution is chromatographed with 2-propanol, ethyl acetate and hydrochloric acid (30 : 10 : 10), in which nickel remains near the start (R_F $0.03 - 0.28$) and zinc and cobalt move faster

(R_F 0.87−0.60). For the determination of nickel in cobalt, and nickel and cobalt in manganese, a mixture of cyclohexanone, 2-propanol, ethyl acetate, ether and hydrochloric acid (30 : 5 : 10 : 20 : 14) is used. R_F values in this solvent mixture are Ni 0.06, Mn 0.26−0.54, Co 0.62−0.88, Zn 0.90, and Fe 0.98. Control chromatograms are detected with potassium ferricyanide and the corresponding zones on the chromatograms of the analysed samples are cut out and eluted as for gold. The determination is photometric, cobalt with nitroso-R salt, zinc with dithizone, and nickel with dimethylglyoxime.

High-purity lead is dissolved in boiling 40% nitric acid and the solution evaporated to dryness, and extracted with water acidified with a few drops of nitric acid. For the determination of bismuth and cadmium a mixture of acetone, water and nitric acid (30 : 5 : 5), or methyl isobutyl ketone, acetone, water and nitric acid (15 : 15 : 5 : 3) is used as solvent. Bismuth and cadmium are then eluted and determined photometrically, bismuth with potassium iodide, and cadmium with dithizone.

To determine zinc and lead in high-purity gallium the sample is dissolved in boiling *aqua regia* and the chromatographic separation is carried out with ethyl acetate, water and nitric acid (30 : 1 : 15). Gallium moves with the solvent front, lead and zinc are slower (R_F 0.12−0.17, and 0.15−0.20). The final determination is done photometrically with dithizone.

When zinc and lead are to be determined in high-purity cadmium, the sample is dissolved in dilute nitric acid or *aqua regia*. For the separation of small amounts of lead, a mixture of ethyl acetate, isopentyl acetate, hydrochloric acid and cyclohexanol (20 : 10 : 12 : 4) is suitable and for the separation of copper, a 1-propanol, chloroform, hydrochloric acid and water mixture (15 : 15 : 1.5 : 0.2). In both systems the cadmium zone is preceded by the zone of the elements to be determined.

For the determination of copper in high-purity zinc the sample is dissolved in dilute *aqua regia* and chromatographed with isobutyl alcohol, chloroform, hydrochloric acid and water (18 : 12 : 1 : 1). The clearly separated zone of zinc moves with the solvent front while copper remains near the start.

For the determination of gold and silver in copper [67], 0.5 g of sample is dissolved in 3 ml each of $8M$ hydrochloric acid and concentrated nitric acid, the solution is evaporated to dryness twice with $0.1M$ nitric acid, and the residue is dissolved in $0.1M$ hydrochloric acid. Gold(III) and silver (about 100 µg) are adsorbed by 25 ml of the solution passing 7 and 9 times, respectively, at 6−8 ml/min through paper discs (18 mm diameter, 0.33 mm thick) impregnated with Amberlite SB-2. The discs are then placed in a sample holder made of polyvinyl chloride and submitted to X-ray analysis utilizing the Au $L\beta_2$ and Ag $K\alpha$ lines.

For determination of the composition of cobalt-zinc ferrites [68], a solution of the ferrite, containing 0.78−3.8 µg of iron, 0.12−1.65 µg of cobalt and 0.13−1.87 µg of zinc in 2.11 µl of dilute nitric acid (pH 1−2), was applied to a strip of filter paper (5 cm × 7 mm) that had been impregnated with $0.25M$ sodium fluoride and then passed through $0.5N$ barium nitrate. This strip was joined at the top to a second strip (3 cm × 8 mm) impregnated with barium sulphate and then with 0.2% 1-nitroso-2-naphthol solution in 80% acetic acid. In turn this strip was joined to a third (7 cm × 1 cm) impregnated with $0.0075M$ potassium ferrocyanide and then with uranyl nitrate solution. The chromatogram was developed by the ascending technique with water -80% acetic acid (193 : 7). Iron was detected on the lowest strip by dipping the paper in $0.25M$ potassium ferrocyanide acidified with hydrochloric acid, cobalt gave a red-brown spot on the middle strip, and zinc was located on the third strip by dipping in dithizone solution. To determine the elements the areas of the spots were measured.

7.1.3. Analysis of Water

Determination of Aluminium [25]

From a suitable amount of water silica is removed in the usual manner by means of hydrochloric acid. Hydrofluoric acid is added to the crude silica and then evaporated. The residue is fused with potassium bisulphate and the cooled melt extracted with hydrochloric acid. The extract is combined with the filtrate from the silica separation and diluted with water to 150 ml. If iron is not present, 1 mg of iron is added as ferric chloride. The mixture is heated to boiling, cooled, and ammonia is added to neutralize most of the acid. Ammonium chloride is then added, followed by more ammonia until the solution is alkaline to Methyl Red. The mixture is boiled briefly and filtered while hot. The precipitate is washed several times with hot 2% ammonium nitrate solution containing several drops of ammonia and dissolved in dilute hydrochloric acid. The solution is concentrated to a small volume (5 $-$ 10 ml) and a portion containing approximately 5 µg of aluminium is transferred to a strip of chromatographic paper and developed with a mixture of 45 ml of acetone, 45 ml of ethyl acetate, 5 ml of water and 5 ml of hydrochloric acid. Aluminium remains on the start while iron moves with the solvent front. The aluminium zone is cut out and eluted with 1% hydrochloric acid overnight. Testing the eluted segment of paper with Aluminon or morin indicates whether the aluminium is completely washed out from the paper. Aluminium in the eluate is determined photometrically with Aluminon at pH 4.2.

Determination of Lead [26]

A suitable amount of water is evaporated to dryness and the residue extracted with 0.5 ml of dilute hydrochloric acid (1 : 1). A 0.03-ml portion of the clear solution is put on a chromatographic paper previously washed with a mixture of acetone and $0.1M$ hydrochloric acid (2 : 1) for 24 hours and dried in air. Chromatography is carried out with a mixture of acetone, isobutyl alcohol, $6M$ hydrochloric acid and acetylacetone (30 : 4 : 2 : 1) for 2 $-$ 3 hours. The chromatogram is dried in air. A control chromatogram is detected with ammonia and dithizone. The R_F values are 0.95 for zinc, 0.79 for Fe, 0.50 for Cu, and 0.25 for lead. The lead zone on the undetected chromatogram is cut out and eluted. The lead in the eluate is determined photometrically with dithizone by extraction at pH 9.4 into chloroform.

Determination of Zinc [27]

A 100-ml sample of water is evaporated to dryness on a water-bath and the residue extracted with 0.5 ml of dilute hydrochloric acid. An aliquot is transferred to a strip of chromatographic paper and developed along with a control sample with a mixture of isobutyl alcohol, acetone, acetic acid, hydrochloric acid and water (16.7 : 64.7 : 3 : 0.9 : 12.4). The chromatogram is dried and the control chromatogram cut longitudinally into two parts, one of which is detected with a solution of dithizone in chloroform (Zn gives a red spot) and the second with a solution of Aluminon (Fe gives a violet spot). The zone on the chromatogram of the analysed sample corresponding to zinc is cut out and zinc is eluted from the paper with a solution of dithizone in chloroform. The eluate is mixed with 5 ml of 0.1% dithizone solution in chloroform and 5 ml of buffer (pH 8.2; 50 ml of $0.1M$ boric acid in $0.1M$ potassium chloride, 5.9 ml of $0.1M$ sodium hydroxide, and 44.1 ml of redistilled water), and shaken for a few minutes. The intensity of the colour formed is measured photometrically.

Determination of Chlorides, Bromides and Iodides

Gantchev and Koev [28] proposed a method for the determination of chloride, bromide and iodide in natural water, based on the separation of the ions on a paper impregnated with silver chloride or chromate. The method permits the determination of $5 - 50$ µg of chloride, $10 - 80$ µg of bromide, and $15 - 100$ µg of iodide with an error of $\pm 1\%$.

7.1.4. Other Applications

In quantitative analysis paper chromatography may be applied to a wide variety of materials. A discussion of the separation of inorganic components in organic materials has been omitted intentionally, as it does not fit into the scheme of this monograph. However, for the sake of completeness it should be mentioned that paper chromatography has found wide application here too. For example, methods have been described for the separation of inorganic poisons (for example thallium) in toxicological analyses, for the determination of trace elements (for example copper in blood), arsenic in biological material, calcium and magnesium in milk, etc. Similarly, procedures developed for the analysis of typical inorganic materials will not be given in detail, but the analysis of coal ash, iron oxides and ferrites, and commercial tripolyphosphates will be quoted as examples. Most of the separations described in Chap. 6 were intended originally for use in applied analysis, or are readily adaptable for such purposes.

Determination of Scandium in Coal Ash [29]

The ash (0.5 g) is dissolved in a mixture of sulphuric and hydrofluoric acids and the solution evaporated to dryness. The residue is wetted with water and evaporated again. It is then leached with water and filtered if necessary. The residue is ignited, fused with potassium bisulphate, dissolved in water, and the solution combined with the filtrate, which is then neutralized with 5 – 10% sodium hydroxide solution (Congo Red indicator), boiled briefly and mixed with hot sodium hydroxide to give a final alkali concentration of 1 – 1.5%. The residue is filtered off and washed on the filter 3 – 5 times with hot water. If iron is absent, approximately 20 mg of iron as ferric chloride are added before precipitation of the hydroxides. If the ash contains less than 20% of aluminium, the precipitation with sodium hydroxide is not necessary.

The precipitate on the filter is dissolved in hot hydrochloric acid and to the solution are added 25 ml of 40% tartaric acid solution, 25 mg of yttrium oxide dissolved in hydrochloric acid and a few drops of Phenol Red indicator. The solution is diluted to 100 ml with water, neutralized in the cold with ammonia, boiled for 3 minutes, mixed with 5 ml of 25% ammonia solution and allowed to stand overnight. The precipitate of basic tartrates is filtered off and washed with 4% tartaric acid solution neutralized with ammonia (Phenol Red indicator). The filter with the precipitate is ignited at 600° and the residue in the crucible dissolved in 2 ml of $6M$ hydrochloric acid, the solution is evaporated to dryness, and the residue is dissolved in 0.25 ml of $4M$ hydrochloric acid. An aliquot (0.1 ml) is applied to a chromatographic paper strip, 26 cm long and 2 cm broad, previously impregnated with 10% ammonium nitrate solution and dried in air. Development is with the organic phase of a thoroughly shaken mixture of 100 ml of iso-pentanol, 100 ml of dilute hydrochloric acid (1 : 5), 100 ml of 30% ammonium thiocyanate solution, 25 ml of benzene and 25 ml of acetone. After about 20 hours of chromatography scandium has moved 15 – 18 cm, while yttrium remains on the start. When the separation is ended the chromatogram and the control chromatogram (run alongside) are dried in air and scandium on the control chromatogram is detected with 0.5% arsenazo I solution and exposure to ammonia vapour. The corresponding zone on the analytical chromatogram is cut out and placed in a 25-ml volumetric flask, then 10 drops of $6M$ hydrochloric acid are added, followed by 10 ml of water and 1 ml of a 20% hydroxylamine hydrochloride solution The flask is heated for 3 – 5 minutes in a boiling water-bath and after cooling 0.5 ml of a 0.1% aqueous alizarin solution is added, followed by 10% ammonia solution until the colour changes. The mixture in the flask is then buffered with 2.5 ml of ammonium acetate buffer of pH 3.5 and made up with water to the mark. The scandium content is determined photometrically.

Analysis of Iron Oxides

The use of paper chromatography in the determination of trace impurities in iron oxides and ferrites was proposed by Bovalini and Piazzi [30]. They determined down to 0.005% of aluminium, titanium, manganese, cobalt, copper or zinc in the samples, by the following procedure. A 5-g sample is ignited at 450 – 500° and then decomposed by repeated evaporation with concentrated hydrochloric acid. The residue is extracted with a small amount of $6M$ hydrochloric acid and portions of the extract are applied to several strips of chromatographic paper and developed with 1-butanol, ethyl acetate and hydrochloric acid (45 : 45 : 10). The chromatograms are dried and detected with morin, dithizone, or 1-nitroso-2-naphthol. The zones of the corresponding metal ions are evaluated by direct comparison with spots on standard chromatograms.

Analysis of Commercial Tripolyphosphates

Gauthier [31] devised a method for the determination of orthophosphate in commercial tripolyphosphates. Approximately 0.5 g of sample is dissolved in 100 ml of water. An aliquot (0.5 ml) is applied to a chromatographic paper and developed with a mixture of 150 ml of dioxan, 50 ml of 2-propanol, 100 ml of water and 0.5 g of boric acid, adjusted with sodium hydroxide to pH 7.2 — 7.3. A control chromatogram run alongside is detected with molybdate, to give phosphomolybdenum blue, and the corresponding zone is cut out of the chromatogram of the sample and extracted in a small extractor with 50 ml of water. The R_F value of orthophosphate is 0.4, while diphosphate and tripolyphosphate remain near the start. The extract is concentrated to 10 ml and mixed with 20 ml of the reagent solution (25 ml of 2% sodium molybdate in $5M$ sulphuric acid, 65 ml of water and 20 ml of a mixture prepared by mixing 5 ml of 60% perchloric acid, 10 ml of $0.1M$ hydrochloric acid, and 1 g of ammonium molybdate, and made up to 100 ml with water. The mixture is transferred to a 50-ml standard flask and diluted to the mark with water. The flask is placed in a boiling water-bath for 10 minutes, then cooled, and the absorbance is measured at 830 nm.

7.2. Quantitative Chromatography on Cellulose Columns

In contrast to paper chromatography, which is suitable for work on the microscale and the analysis of very small amounts of material, column chromatography permits the analysis of larger amounts of sample. This is usually advantageous if minor components have to be determined and large amounts of the material are available, as with ores, metals and alloys. For the analysis of ores, methods have been developed mainly by workers at the Chemical Research Laboratory, Teddington, England, who concentrated their efforts mainly on the determination of uranium, thorium, niobium and tantalum.

7.2.1. Analysis of Ores

Determination of Uranium in Silicates [32]

A finely ground sample (2.5 g) is decomposed in a platinum dish with 2 ml of concentrated nitric acid and 5 ml of hydrofluoric acid and the solution evaporated almost to dryness. The mixture is washed into a 100-ml beaker with a minimum amount of water and evaporated to dryness. The residue is treated with 10 ml of concentrated nitric acid and again evaporated to dryness. The residue is extracted with 8 ml of water and 2 ml of nitric acid.

Cellulose (450 g) for chromatography is boiled for 2 minutes with 3 litres of 5% nitric acid. It is then filtered off and washed with water until neutral, then with ethanol, and finally with ether, and dried under suction. A glass tube of 2 cm in diameter and 40 cm in length is made water-repellent by rinsing with dichlorodimethylsilane, and is three-quarters filled with ether containing 5% (v/v) concentrated nitric acid, then cellulose is added gradually in small portions until it occupies half the tube. The column is then washed with about 100 ml of ether containing 5% (v/v) concentrated nitric acid and the solvent is allowed to drain just to the level of the column.

The sample is mixed with enough cellulose powder to imbibe it all, then the mixture is homogenized by stirring with a glass rod, and introduced on to the top of the column. The remaining slurry in the walls of the beaker is rinsed onto the column with a maximum of 10 ml of ether containing 5% (v/v) concentrated nitric acid. The top layer of the column is stirred and allowed to settle. It is then pressed gently and a 250-ml beaker is placed under the column and the solvent is allowed just to soak into the column. The column is then washed with 10-ml portions of the solvent which are used to rinse the beaker in which the sample was homogenized. After 100 ml of eluate have been collected, elution is stopped.

Water (50 ml) is then added to the eluate and the organic phase is evaporated on a water-bath. The aqueous residue is mixed with 5 ml of sulphuric acid and 5 ml of perchloric acid and evaporated until dense white fumes are evolved. Uranium is then determined polarographically or photometrically. Higher concentrations can be titrated.

Determination of Uranium in Minerals and Ores Containing Arsenic and Molybdenum

Ryan and Williams [33] proposed a suitable method of chromatographic separation of uranium for materials also containing arsenic and molybdenum. The latter two elements are not bound on cellulose when chromatography is carried out with ether containing 5% nitric acid and they pass into the eluate where they might interfere with the determination of uranium. These authors therefore proposed the use of a column containing both cellulose and alumina. The latter does not adsorb uranium from the solvent, but it does adsorb arsenic and molybdenum.

The column is prepared as follows. A glass tube 2.7 cm in diameter and 25 cm long is coated inside with a silicone to make it water-repellent and then three-quarters filled with ether containing 5% (v/v) of concentrated nitric acid. Cellulose powder (5−6 g) is then introduced to make a column and 15 g of alumina are then added. The double column is then washed with the solvent and is ready for the separation.

The sample (100−150 mg of U_3O_8) is weighed in a platinum crucible, and decomposed with a mixture of nitric and hydrofluoric acids. The hydrofluoric acid is then removed by repeated evaporation with nitric acid. If the sample is not decomposed completely the undissolved residue is filtered off, ignited, then fused with a few pellets of potassium hydroxide in a nickel crucible, and the cooled melt is extracted with the filtrate. The solution is evaporated to dryness, and the residue dissolved in 4 ml of dilute nitric acid (1 : 3) by gentle heating. After cooling, the solution is completely imbibed on added cellulose, and the whole transferred onto the column in the manner just described. If arsenic and molybdenum are present, 200 – 250 ml of the solvent are required for elution; if only molybdenum is present 150 ml should suffice. Water is added to the eluate and the organic phase evaporated on a water-bath. The residue is wetted with sulphuric and perchloric acids and evaporated until white fumes appear. The residue is extracted with water, and uranium is then determined titrimetrically.

Determination of Uranium and Thorium in Uranothorianite

If a suitable concentration of nitric acid in ether is chosen for the eluent, uranium and thorium can also be separated. While uranium is eluted from the column when the concentration of nitric acid is only 3% (v/v), thorium is retained by the column. However, thorium can be eluted with ether containing 12.5% (v/v) of concentrated nitric acid. This effect was used for the determination of both elements in uranothorianite [34].

A finely ground sample (0.4 – 0.05 g) is weighed into a 100-ml beaker and mixed with 10 ml of concentrated nitric acid. The beaker is covered with a watch-glass and digested on a water-bath for 2 hours. Water (5 ml) is added and if an insoluble residue is present it is filtered off. The filter is ignited, fused with a few pellets of potassium hydroxide, extracted with the filtrate, and the whole evaporated to dryness. The residue is extracted with 5 ml of water and evaporated with nitric acid with some hydrogen peroxide present. The residue is extracted with 2 ml of dilute nitric acid and then transferred, after mixing it with cellulose as above, onto a cellulose column prepared in ether containing 5% (v/v) of concentrated nitric acid, and eluted with 150 ml of the same solvent mixture. The elution is continued with 600 ml of ether containing 12.5% (v/v) of concentrated nitric acid. The organic solvent is eliminated from both fractions by evaporation on a water-bath. The uranium in the first fraction is determined titrimetrically, the thorium in the second fraction is determined gravimetrically with oxalic acid.

Simultaneous Determination of Thorium and Uranium in Minerals and Ores

For a simultaneous determination of thorium and uranium, Williams proposed [35] a combined column of cellulose and alumina. The column is eluted first with ether containing 1% (v/v) of concentrated nitric acid to elute the uranium, and then with ether containing 12.5% (v/v) of nitric acid to wash out the thorium. The method is suitable both for the analysis of simple minerals, such as monazite, and for the analysis of complex materials, such as samarskite or pyrochlor, which are difficult to analyse by other methods.

Potassium hydroxide (10 g) is weighed in a nickel crucible, gently heated to eliminate water, and cooled. Then 1 g of sample is added, the crucible is covered, and gently heated to red heat for about 1 hour. The melt is cooled and extracted with 150 ml of dilute nitric acid (1 : 3). The crucible is rinsed with water and the solution boiled for 5—10 minutes. A completely clear solution can be sometimes obtained by adding one drop of dilute hydrofluoric acid (1 : 100), but not if the sample is rich in zirconium. The solution is cooled a little, made slightly alkaline with ammonia, filtered, and the precipitate washed twice with hot water containing a little ammonia. The precipitate is then washed off the filter into the beaker with 50—100 ml of hot water. Concentrated nitric acid (40 ml) is added and the mixture evaporated to yield a moist residue, which is then treated with 20 ml of dilute nitric acid (1 : 3) and the beaker is covered and gently heated for 5 minutes. The beaker is then uncovered, hydrogen peroxide (2 ml) is added and the mixture heated for 10 minutes to reduce any cerium salts. Ferric nitrate (8 g) is added and the mixture heated for 15 minutes, then 1.4 g of disodium hydrogen phosphate are added and the mixture heated for another 15 minutes, and then cooled. Next 50 g of alumina are added, the mixture is homogenized by stirring, and introduced onto a column of cellulose and alumina, prepared in ether containing 1% (v/v) of concentrated nitric acid, and uranium is eluted completely with 400 ml of the same solvent, and can be determined in the eluate (after evaporation of the solvent) by photometry, polarography or titrimetry. Thorium is then eluted with ether containing 12.5% nitric acid (v/v) and determined gravimetrically with oxalic acid.

Determination of Thorium in Monazite [34]

A sample containing at most 250 mg of thorium oxide is weighed into a platinum crucible, 10 ml of hydrofluoric acid are added, and the mixture is evaporated on a steam-bath almost to dryness. The residue is mixed with 20 ml of 5% hydrofluoric acid, heated, and the solution removed through a polyethylene filter-stick. This step is repeated twice more, and after final removal of the solution the precipitate is rinsed into a nickel crucible. The filtrate from the filter-stick is put into the crucible and dried on a water-bath. Potassium hydroxide (5 times the sample weight) is then added and heated under an infrared lamp for 15 minutes to eliminate moisture. The mixture is then fused for 45 minutes. The melt is cooled and digested with 200 ml of water in a 250-ml beaker. The crucible is rinsed and a little soluble starch is added to the solution which is then boiled. After cooling and settling, the clear supernatant liquid is filtered off with a filter-stick. A fresh 200-ml portion of water is added and boiled, and the supernatant filtered off as above after settling of the precipitate. The filter paper from the stick is decomposed by digestion with 1 ml of fuming nitric acid, and the product transferred to the main beaker. Nitric acid (20 ml) is added and the mixture gently boiled. The solution should be clear. If not, the attack with nitric acid is repeated. Water (5 ml) is then added, followed by 0.2 ml of nitric acid and 0.5 ml of hydrogen peroxide, and the solution is heated on a steam-bath until the red colour completely disappears. It is then cooled, 2 ml of nitric acid are added and a suitable amount of cellulose to transfer the material onto a cellulose column prepared in ether containing 12% (v/v) nitric acid. Thorium is eluted with 600 ml of the same solvent. To the eluate, 100 ml of water containing

30 ml of ammonia are added and the ether is distilled off. The residue is neutralized with ammonia, hydrochloric acid (25 ml) is added and the mixture is diluted with water to 400 ml. The solution is gently boiled, 20 g of oxalic acid are added in small portions with stirring, and the mixture is cooled and filtered. The beaker and the material on the filter are washed with 100 ml of oxalic acid solution in $0.1M$ hydrochloric acid. The filter is then dried, and ignited at 850°, and the product weighed as thorium oxide.

The following procedure may also be used for the decomposition of monazite. The sample is dissolved in hot sulphuric acid. (If the monazite is not pure it is usually necessary to fuse the insoluble residue with potassium hydroxide.) The acidity of the solution is then adjusted with hydrochloric or sulphuric acid to $0.5N$, the solution is diluted to 200 ml, gently boiled with continuous stirring, and the oxalates precipitated with 100 ml of boiling oxalic acid. After cooling, the precipitate is filtered off, washed, rinsed into a 100-ml beaker, and evaporated to dryness. Fuming nitric acid (20 ml) is added and evaporated, and the operation repeated to ensure a complete decomposition of the oxalates. The residue is dissolved in water and boiled with nitric acid and hydrogen peroxide as in the first method, the procedure of which is then followed.

For low concentrations of thorium in minerals and ores, Kingsbury and Temple [36] proposed a separation on a combined cellulose and alumina column (see p. 169). Ether containing 12.5% (v/v) of nitric acid is used as eluent. The extracted thorium is precipitated together with lanthanum which serves as an internal standard during the final spectrophotometric determination. By this method $0.0001-0.2\%$ of thorium can be determined with $\pm 3\%$ error.

Determination of Niobium and Tantalum in Minerals and Ores

For the separation of niobium from tantalum and for their determination in certain types of artificially prepared or natural mixtures, a cellulose column is suitable with a mixture of ethyl methyl ketone and hydrochloric acid for elution. Williams and Burstall [37, 38] applied this procedure for the quantitative determination of niobium and tantalum in complex materials containing appreciable amounts of titanium, tin and zirconium, especially the minerals columbite and pyrochlor, and iron, tantalum, niobium alloys.

The material is decomposed by evaporation with hydrofluoric acid and nitric acid in a platinum dish. After the decomposition the residue is evaporated several times with hydrofluoric acid to transform all the material into fluorides. The residue is dissolved in 8 ml of 25% hydrofluoric acid and heated gently in a covered dish (complete evaporation should be avoided). Ammonium fluoride (1 g) is added and the heating is continued for a few minutes. The solution is cooled, 6 g of cellulose are added and the mixture is homogenized and transferred onto a cellulose column prepared in ethyl methyl ketone saturated with water (400 ml of ethyl methyl ketone shaken with 50 ml of water). The same solvent is used for the elution until 250 ml of the eluate are collected. This fraction contains all the tantalum.

Titanium, tin and zirconium are then eluted with 400 ml of ethyl methyl ketone containing 1% (v/v) of concentrated hydrofluoric acid. In the case of certain materials the last 200 ml of eluate may already contain a certain amount of niobium. This must be checked by a suitable reaction. If niobium is present in the eluate the last 200 ml of it are combined with the further eluate obtained with 400 ml of ethyl methyl ketone containing 12.5% (v/v) concentrated hydrofluoric acid (this fraction contains all the niobium).

The solvent is evaporated from the eluate on a water-bath and the aqueous residue transferred quantitatively into a special porcelain dish, 2 ml of 50% sulphuric acid are added, and the solution is concentrated by heating under an infrared lamp, to a volume of 1 ml. A piece of filter paper is then added to the dish to soak up all the solution, and is then dried and finally ignited at 900°. The oxide formed (Ta_2O_5 or Nb_2O_5) is then weighed. If the original material contains tungsten it is necessary to find out whether this is also present in the niobium oxide. On fusing with sodium carbonate and extraction with hydrochloric acid, tungsten passes into solution and can be determined photometrically with thiocyanate.

Mercer and Williams [39] extended this method to the determination of niobium and tantalum at concentrations lower than 0.1%; in this case the sample must be pretreated. The method is suitable mainly for the determination of niobium and tantalum in phosphate ores.

The sample (10 g) is decomposed by boiling with 100 ml of 18% hydrochloric acid, then evaporated to dryness; a small amount of hydrochloric acid is added and the evaporation repeated. The residue is heated under a infrared lamp for one hour in order to eliminate fluorides which might be present in the sample. The residue is extracted with 5 ml of concentrated hydrochloric acid and 250 ml of water and mixed with 30 ml of a saturated solution of sulphur dioxide. The solution is boiled for 20 minutes and allowed to stand overnight. It is then filtered, the filter is washed with hot $0.5M$ hydrochloric acid, followed by 100 ml of 18% hydrochloric acid, and eventually with 30 ml of hydrogen peroxide. The filter is ignited in a platinum dish, some of the wash solution is added and the mixture heated for a few minutes. The solution is diluted to 600 ml and boiled for 30 minutes, the volume of the solution being kept above 500 ml, allowed to stand for 1 hour and filtered. The filter is washed with 0.5% hydrochloric acid, and retained. The filtrate is neutralized with ammonia, then 10 ml of concentrated hydrochloric acid and 30 ml saturated solution of sulphur dioxide are added. The solution is boiled for 30 minutes and allowed to stand overnight. After filtration the precipitate is combined with the first precipitate in a platinum dish, and ignited. The residue is composed mainly of niobium pentoxide, with some silica, calcium phosphate and phosphates of some other metals (titanium, zirconium). The residue is dissolved in hydrofluoric acid and chromatographed as before.

Mercer and Wells [40] conveniently shortened this procedure. Their method is a simultaneous chromatographic extraction of niobium, tantalum and tungsten with ethyl methyl ketone containing 15% (v/v) of hydrofluoric acid, and another chromatographic separation or a direct determination in the first eluate by suitable photometric methods.

The sample is fused with bisulphate, the melt is extracted with 200 ml of 5% sulphuric acid and boiled until the melt is decomposed completely.

Hydroxides are precipitated from the hot solution with ammonia, the precipitate is filtered off and washed with 2% ammonia—5% ammonium nitrate solution. The filter with the precipitate is ignited in a platinum dish at $500-600°$ and the residue digested with a mixture of hydrofluoric and nitric acids. Nitric acid is removed by repeated evaporation with hydrofluoric acid. The residue is extracted with 4.5 ml of water with 1 g of ammonium fluoride and 1.5 ml of hydrofluoric acid added and is gently heated for a few minutes until dissolution is complete. However, care must be taken to avoid evaporation of the acid. The solution is cooled, mixed thoroughly with approximately 5 g of cellulose powder, transferred onto a cellulose column prepared with ethyl methyl ketone containing 15% (v/v) of concentrated hydrofluoric acid, and eluted with 400 ml of the same solvent. The eluate is evaporated on a steam-bath and the residue is transferred to a platinum dish and evaporated almost to dryness under an infrared lamp. Concentrated sulphuric acid (5 ml) is added, followed by a few drops of nitric acid, and the mixture is evaporated until dense white fumes are evolved. The residue is diluted with water to 250 ml, boiled, and neutralized with ammonia. An additional 10 ml of ammonia are added and the precipitate is digested for 30 minutes on a warm bath. After filtering and washing with hot 5% ammonium nitrate-2% ammonia solution, the filter with the precipitate is transferred into a weighed platinum dish and ignited. The residue contains the oxides of niobium and tantalum. However, if the original solution contained much phosphorus the ignited oxides should be redissolved in hydrofluoric acid, evaporated to white fumes after the addition of sulphuric acid, extracted, and the precipitation with ammonia repeated. In the mixture of oxides niobium and tantalum are then determined colorimetrically.

7.2.2. Analysis of Metals and Alloys

Systematic Analysis of Special Steels

Venturello and Ghe [41] described a chromatographic separation and determination of the most important elements, including molybdenum, cobalt, manganese, vanadium, nickel and chromium in special steels. The method is based on the photometric determination of elements eluted from a cellulose column, and it permits the determination of micro amounts with an absolute error of $0.02-0.05\%$.

A sample of the steel (0.25 g) is decomposed with 2 ml of $6M$ hydrochloric acid by heating on a sand-bath. During the decomposition hydrochloric acid is added dropwise until the volume is 8 ml. The solution is evaporated to small volume, 3 ml of $6M$ hydrochloric acid are added, the mixture is warmed, and a further 3 ml of hydrochloric acid are added dropwise. The hot solution is oxidized with 6 drops of concentrated nitric acid, evaporated almost to dryness, extracted with 2 ml of $6M$ hydrochloric acid, and diluted to volume in a 5-ml volumetric flask.

An aliquot (0.2 ml) is transferred onto a cellulose column 5 mm in diameter and 11.5 cm long, prepared by filling the tube with dry cellulose, pressing, and wetting with acetylacetone. The elements are eluted as descri-

Table 7.1. Fractional Elution of Elements from a Cellulose Column
According to Venturello and Ghe

Solvent used	Amount ml	On the column	In the eluate
acetylacetone	7	Co, Mn, V, Ni, Cr	Fe, Mo
methyl *n*-propyl ketone containing 5% of HCl	7	V, Ni, Cr	Co, Mn
methyl *n*-propyl ketone containing 5% of HCl	13	Ni, Cr	V
a mixture of water and dilute H_2SO_4 (1 : 20)	8 + 2	—	Ni, Cr

bed in Table 7.1. In the eluate molybdenum is determined with potassium
thiocyanate after the reduction of iron(III) to iron(II) with stannous chloride,
cobalt is also determined with thiocyanate (the reaction is not affected by
the presence of manganese), manganese is determined by its oxidation with
persulphate to permanganate, vanadium as tungstovanadophosphate, nickel
with dimethylglyoxime in the presence of an oxidizing agent and with masking of chromium with citric acid, and chromium is determined with diphenylcarbazide.

Determination of Aluminium in Iron and Steel [42]

A steel sample containing about 50 µg of aluminium is dissolved in 10 ml
of concentrated hydrochloric acid. Water (10 ml) is added and the mixture
oxidized by dropwise addition of 1 ml of hydrogen peroxide (100 vol).
After gentle boiling to eliminate excess of peroxide, the mixture is evaporated to 5 ml, and cooled. A suitable amount (usually 25 − 50 ml) of freshly
distilled ethyl methyl ketone is added, and if necessary, concentrated hydrochloric acid to make the solution homogeneous.

The column is prepared by mixing 5 g of cellulose with 90 ml of hydrochloric acid (1 : 5), introducing the mixture into the tube, washing with an
additional 200 ml of acid and then 200 ml of water, followed by 100 ml of
freshly distilled ethyl methyl ketone and finally with 100 ml of ethyl methyl
ketone containing 4% of hydrochloric acid. This operation is necessary for
the elimination of aluminium, if present in the cellulose.

The sample solution is transferred onto the column with 2-ml portions
of the ketone eluent until all the iron has been washed out of the column.
The column is washed with a further 50 ml of the solvent. The eluate is
discarded and aluminium (together with any nickel) is eluted with 200 ml
of hydrochloric acid (1 : 5). The eluate is evaporated to dryness, 0.5 ml of
60% perchloric acid is added and the mixture evaporated to dryness. The
residue is dissolved in a few drops of dilute perchloric acid (1 : 1), 2 drops

of Methyl Red solution are added, and the mixture rendered alkaline with 10% sodium hydroxide solution. It is then acidified with $1M$ perchloric acid and then mixed with 0.2 ml of $5M$ perchloric acid, 1 ml of sodium acetate, and 5 ml of 0.1% Solochrome Violet RS solution, transferred into a 10-ml volumetric flask and diluted to volume with water. The flask is immersed in a water-bath at $55-70°$ for 5 minutes, cooled, de-aerated and polarographed between 0 and -0.8 V *vs.* a saturated calomel electrode.

Small amounts of nickel do not interfere with the determination. If more nickel is present it must be eliminated beforehand by electrolysis on a mercury cathode.

Determination of Gold

By use of a cellulose column and a mixture of ethyl acetate and nitric acid as solvent, gold can be successfully separated from the platinum metals [43].

The sample (0.25 g) is dissolved in *aqua regia* and the solution is evaporated on a steam-bath, to a syrup. Hydrochloric acid (5 ml) is added and evaporated to dryness. The residue is mixed with 10 ml of $2M$ hydrochloric acid and 0.5 g of sodium chloride and evaporated almost to dryness. The sodium salts are dissolved in 1 ml of concentrated hydrochloric acid saturated with chlorine, and 4 ml of water.

The column is prepared from a suspension of cellulose in ethyl acetate containing 2% (v/v) of concentrated nitric acid and 3% (v/v) of water. The height of the column should be 7 cm. Then 2 g of cellulose powder are added to the sample solution and the mixture is homogenized and transferred onto the column with small amounts of solvent, and eluted at a flow-rate of 2.2 ml/min with 100 ml of solvent used in 10-ml portions to rinse out the beaker. The eluate is collected in a 500-ml conical flask, water (50 ml) is added and the solvent is evaporated. Hydrochloric acid (5 ml) is added to the residue and evaporated to small volume. The residual solution is rinsed with $2M$ hydrochloric acid into a suitable electrolytic vessel. After heating to $50°$ the solution is stirred with a stream of hydrogen or air and electrolysed at $1.5-3$ mA/cm^2 current density at $2.5-3.5$ V for 30 minutes. The cathode is rinsed with water and then with acetone, and dried at $110°$, and the separated gold is weighed. The electrolysis is repeated for 15 minutes each time until the weight of the separated gold is constant.

Determination of Zinc in Copper Alloys

Chew and Lindley [44] described a method for separation of zinc suitable for all types of alloys based on copper with a content of $1-40\%$ of zinc.

A tube 2 cm in diameter and 15 cm in length is filled with cellulose suspended in a solvent composed of 20% hydrochloric acid and 1-butanol (4 : 96), and then washed with 100 ml of the solvent.

A sample (0.1 g) is dissolved in *aqua regia*, 1 ml of 5% ferric chloride solution is added and the solution is evaporated almost to dryness. The residue is dissolved in a minimum amount of solvent and mixed with as much cellulose as necessary to imbibe the whole solution. After thorough stirring the mixture is transferred onto the column and rinsed with 10 ml of solvent. The upper part of the column is stirred and pressed with a rod. The solvent is allowed to drain until all the solvent above the column has just soaked into it. The beaker and wall of the tube are rinsed with 10 ml of solvent, then elution is continued until the yellow zone of iron has passed into the eluate. The eluate is evaporated to $10-20$ ml, 50 ml of a buffer solution (80 ml of ammonia and 10 g of ammonium chloride per litre) are added and the solution is titrated with $0.1M$ EDTA solution, with Eriochrome Black T as indicator.

7.3. Quantitative Chromatography on Silica Gel Columns

In organic analysis silica gel is often used for adsorption, partition, and gas chromatography, but in inorganic analysis it is used relatively little. It is difficult to account for the sorption effects of silica gel. In aqueous medium it functions as a weak cation-exchanger. However, its capacity for exchange is rather low (0.01-0.04 meq/g). If it is used as the carrier of the stationary phase in partition chromatography, adsorption plays a far greater role than in partition chromatography on cellulose.

Suitable methods have been devised for using silica gel to retain small amounts of colloidal hydroxide and hydroxo complexes. Řezáč and Roubal [45] have shown that small amounts of colloidal stannic hydroxide can be quantitatively retained on a silica gel column, although an attempt to filter off such amounts on filter paper was unsuccessful. They explained the retention of tin as due to the adsorption of the positively charged stannic hydroxide sol on the silica gel surface, where they supposed the presence of silicic acid ions would cause coagulation of positively-charged hydroxide sols.

Determination of Tin in Tungsten Concentrates

The sample (0.5 g) is fused in an iron crucible with 5 g of sodium peroxide. The cooled melt is extracted with water, a few drops of 3% hydrogen peroxide are added, and the mixture is boiled and filtered. The precipitate is washed with warm 1% sodium hydroxide solution and the filtrate is diluted to volume in a standard flask. To an aliquot of the solution containing $1-5$ mg of tin, 10 ml of $0.1M$ EDTA are added, the mixture is neutralized (Methyl Red) with dilute sulphuric acid, ammonia is added until the solution is at pH 9 (indicator paper) and the solution is diluted to 200 ml. The concentration of ammonium sulphate should be $0.2M$. The solution is then transferred onto a silica gel column 10 cm long and $10-15$ mm in diameter, and of particle size $0.2-0.4$ mm. Before use, the silica gel is purified by extraction with concentrated hydrochloric acid and washed with distilled water. The elution rate should be 5 ml/min. When all the solution has passed through, the column is washed with 200 ml of water, and tin is eluted with 6 ml of concentrated hydrochloric acid and 70 ml of water. Ammonium chloride

(21 g) and 0.25% gelatin solution (1 ml) are added to the eluate and the mixture diluted to volume with water in a 100-ml standard flask. Tin is determined polarographically in the range from -0.2 to -0.8 V and the second wave [reduction of tin(II) to tin(0)] is evaluated.

Tin can be determined in tungstic acid in a similar manner. The sample is dissolved in $1M$ sodium hydroxide and the insoluble residue fused with a small amount of sodium hydroxide.

Determination of Small Amounts of Uranium in Minerals [46]

A finely ground sample containing about 0.3 mg of uranium is decomposed with a mixture of 15 ml of concentrated nitric acid and 9 ml of concentrated hydrochloric acid. If the decomposition is not complete the residue is fused with sodium carbonate, the cooled melt dissolved in hydrochloric acid and the solution combined with the filtrate. This is then evaporated to dryness several times, hydrochloric acid being added each time to the residue, and the residue is wetted with 2 ml of concentrated hydrochloric acid and extracted with hot water. The insoluble residue is filtered off and washed, then tartaric acid (3 g) is added to the filtrate, followed by $20-30$ ml of $0.2M$ EDTA, the solution is boiled briefly, cooled, and 25 ml of the precipitant are added [132 g of ammonium sulphate dissolved in 875 ml of 10% ammonia solution (prepared by dissolving gaseous ammonia in distilled water) and diluted to 1 litre with distilled water]. The solution is then transferred onto a column of neutral silica gel (particle size $0.2-0.4$ mm) purified before use by washing with concentrated hydrochloric acid and redistilled water. The flow-rate is adjusted to 5 ml/min. The column is washed with 100 ml of water and the uranium retained on it is dissolved in $1-5$ ml of concentrated hydrochloric acid. The column is washed with a few 5-ml portions of water at a flow-rate of 1 ml/min. The eluate is evaporated to dryness. The residue is wetted with 0.5 ml of concentrated hydrochloric acid, extracted with water, mixed with 0.2 g of ascorbic acid, and neutralized with sodium hydroxide. The solution is transferred to a 25-ml volumetric flask, 1.25 ml of 70% perchloric acid are added followed by 0.5 ml of 0.075% thymol solution, and the solution is diluted to the mark with water. The uranium is then determined polarographically.

The same principle, i.e. the retention of uranium on a silica gel column, has also been used for a rapid semiquantitative determination of uranium [47]. The method is based on the intense fluorescence of uranium adsorbed on the silica gel surface and it is especially suitable in cases where large series of samples are to be analysed. It is very sensitive, allowing the detection of 1 µg of uranium.

A glass tube 10 cm long and 5 mm in diameter is filled with a $3-4$ cm layer of silica gel, $0.08-0.18$ mm particle size, previously washed with an alkaline (pH 9.5) $0.1M$ solution of ammonium chloride. To an aliquot of sample solution, containing not less than 1 µg of uranium, as much $1M$ ammonium chloride is added as is necessary to make its final concentration $0.1M$, followed by $1-30$ ml of $0.2M$ EDTA. The solution is rendered alkaline with dilute ammonia (pH $8.5-9.5$) and allowed to flow through the column at 2 ml/min. When all the solution has passed through, the fluorescence of uranium adsorbed on the upper part of the column is observed and evaluated semiquantitatively by comparison with a series of standards.

Determination of Trace Amounts of Beryllium in Mineral Water and Minerals [48]

Mineral water, containing $1-10$ μg of beryllium, is acidified with 5 ml of concentrated hydrochloric acid, 1 ml of 30% hydrogen peroxide is added, and the solution is evaporated in a silica dish to dryness. The residue is wetted with $1-3$ ml of concentrated hydrochloric acid, extracted with $80-100$ ml of hot water and the separated silicic acid filtered off. The filtrate is neutralized with dilute ammonia (1 : 3) and after the addition of 30 ml of $0.2M$ EDTA, the pH is adjusted to 6.5 with dilute ammonia (1 : 6). The solution is allowed to flow through the silica gel column at a rate of 3 ml/min. The silica gel should be previously stirred with $0.1M$ ammonium chloride. When all the solution has passed through, the column is washed three times with 25 ml and finally with 50 ml of water. The beryllium retained on the column is dissolved in 5 ml of hydrochloric acid (1 : 1). The acid is allowed just to soak in, then 5 ml more are added and allowed to stay on the column for 5 minutes. The column is then washed with water and 80 ml of the eluate are collected in a 100-ml volumetric flask. The eluate is neutralized with 25% sodium hydroxide solution and 4 ml of $2.5M$ sodium hydroxide are added. After the addition of 2 ml of a stannite* solution and 2 ml of 0.02% morin solution in acetone, the mixture is made up to the mark with water and the fluorescence measured.

The mineral cinvaldite was analysed by this method. A finely ground sample $(0.1-1$ g) was fused with sodium carbonate and silica was removed in the conventional manner. The filtrate was made up to 250 ml in a graduated flask and an aliquot containing $1-10$ μg of beryllium was added to $20-30$ ml of $0.2M$ EDTA and the pH adjusted to 5.5 (pH-meter). The rest of the procedure was that just described.

Determination of Tin in Antimony [49]

The weight of the sample should be chosen so that the tin content is not less than 2 mg.

The weighed sample is mixed with concentrated hydrochloric acid, and an amount of bromine approximately a tenth that of the acid is then added carefully. A vigorous reaction takes place and the solution becomes hot. A sample weighing 10 g dissolves completely within $10-15$ minutes. An excess of the acid or of bromine should be avoided. After complete dissolution the excess of bromine is eliminated by boiling for $2-3$ minutes and ascorbic acid is added until the solution is decolorized. Antimony usually contains only small amounts of silica, which are precipitated at this point and need not be specially eleminated.

---

* A solution of 1.90 g of $SnCl_2 . 2 H_2O$ in 2 ml of water is added dropwise with constant stirring to a solution of 2.40 g of NaOH in 10 ml of water and the solution made up to 50 ml and filtered. It should be prepared immediately before use.

This method of decomposition is advantageous because of its speed, but it requires experience and skill. Sulphuric and nitric acid can also be used for the decomposition, as follows. To the weighed sample, 3 parts of concentrated sulphuric acid and 1 part of concentrated nitric acid are added, and it is heated on a sand-bath until no black particles of undecomposed antimony are visible. The sulphuric acid concentration is decreased by partial evaporation of the solution. To the residue concentrated hydrochloric acid is added carefully and the mixture is diluted with water. A sample weighing 2 g can be decomposed in 2 hours by this method.

To the acid sample solution prepared in one of the ways described, a suitable amount of citric acid (at least a 5-fold excess relative to the antimony present) and $25-50$ ml of $0.1M$ ammonium chloride are added, the mixture is diluted with water and the pH adjusted to 5.5 with sodium hydroxide solution. The solution is passed through a column of silica gel of $0.06-0.29$ mm particle size, previously purified by extraction with hydrochloric acid (1 : 1), washing with water and finally with $0.1M$ ammonium chloride. The flow-rate is 3 ml/min and after all the solution has passed through, the column is washed with 200 ml of $0.5M$ sodium citrate at pH 5.5, and finally with water acidified to pH 5.5 with dilute hydrochloric acid. Tin is then eluted from the column with 10 ml of $6M$ hydrochloric acid into a 100-ml graduated flask containing 15 g of ammonium chloride. The column is washed with 70 ml of distilled water, gelatin is added, and the solution is made up to the mark. Tin is then determined polarographically.

Preparation of Zirconium-free Hafnium

Hansen and Gunnar [50] found that on passing hafnium and zirconium chlorides dissolved in methanol through a silica gel column the hafnium was adsorbed on the column and its content in the zirconium could thus be reduced to 0.1%. They used this fact for preparative purposes.

A solution of 455 g of zirconium chloride is dissolved in 2275 ml of anhydrous methanol and then allowed to stand for 3 hours. After filtration the solution (containing 106 mg of total metal oxide per ml) is allowed to run through a tube 27 mm in diameter, filled with 290 g of silica gel of $28-120$ mesh, purified by washing with nitric acid (1 : 1) and activated at 300° for 4 hours. The flow-rate is 3 ml/min and the composition of the eluate fractions is determined spectrographically.

Other Applications

A number of authors have described other methods of separating inorganic mixtures on silica gel. Usually these methods are based on the differential sorption of metals and their complex ions as a function of the pH of the solution. Often very effective separations are achieved. Vydra developed, during a systematic study of the analytical use of silica gel, a method of selective sorption of a phenanthroline complex of cobalt in the presence of EDTA and citric acid [51] and he applied the separation to the determination of microquantities of cobalt in the presence of lead, manganese, beryllium, aluminium, chromium, tin, antimony, bismuth and other metals. He also developed a separation of trace amounts of cobalt from an excess of iron. The sorption of copper, zinc, cadmium, silver and cobalt ammine complexes has also been investigated [69].

Vydra also achieved a selective separation of cobalt during the study of the sorption of some metal ethylenediamine complexes [52]. The sorption of cobalt from a medium $0.1M$ in ethylenediamine is also used by Doležal [53] for the quantitative sorption of palladium complexes which, after elution with hydrochloric acid, can be determined polarographically.

Bozhevolnov and Karakovskaya [54] investigated the possibility of determining iron(II) by means of 1,10-phenanthroline. From a solution at pH 5.5 − 6 and containing hydroxylamine, iron is sorbed quantitatively on a silica gel column on which 1,10-phenanthroline is bound. The height of the iron zone is proportional to the amount in the sample. Vydra and Marková [55] used this method for the determination of traces of iron in some metals, masking nickel, chromium, molybdenum, tungsten and other metals with EDTA and citric acid.

The adsorption of silver can also be utilized for a selective separation of silver from a series of other ions from a medium containing EDTA and citric acid at pH 8.5 or higher [56].

The behaviour and differing affinity of bi- and tervalent ions for silica gel was investigated by Kohlschütter [57, 58]. He demonstrated that aquo-complexes of metals are not adsorbed by silica gel, whereas hydroxy-complexes have a certain sorption affinity for silica gel. On this he based a procedure for the separation of aluminium from magnesium, calcium, barium, zinc, nickel, manganese and copper, and the separation of iron(III) from iron (II) copper, nickel and manganese.

Kuteinikov and Brodskaya [59] investigated the possibility of separating rare earths on a silica gel column. They found that at pH 6 rare earths are quantitatively eluted, while zirconium, titanium, thorium, scandium, iron, aluminium and uranium are retained on the column. They made use of this mainly for the separation of thorium and scandium from rare earths.

The sorption of silica gel was also used for the analysis of the fission products of uranium. Rydberg [60] purified plutonium from zirconium and niobium by sorption on silica gel. Ahrland and co-workers [61] separated plutonium from the fission products of uranium by its sorption on silica gel from a medium containing $0.1 - 0.5M$ nitric acid.

A method of determining small amounts of copper in steel, crude iron, aluminium, nickel or zinc [62] does not use silica gel but is related to this group of separations and is of importance in practical industrial analysis. It is based on precipitation chromatography for which quartz sand coated with cadmium sulphide is used. Cadmium sulphide is formed by precipitation from cadmium acetate. When solutions containing cupric ions are filtered through such a column a black zone of cupric sulphide is formed at the top of the column, and the height of the zone is proportional to the copper content of the sample.

REFERENCES

1. AGRINIER, H. *Commisariat à l'energie atomique, Rapport CEA* No. 2099.
2. AGRINIER, H. *Compt. Rend.* **255**, 2801 (1962).
3. ADER, D. and ALON, A. *Analyst* **86**, 125 (1961).
4. AGRINIER, H. *Chim. Anal. Paris* **42**, 600 (1960).
5. DEMETRESCU, A. and CIOACA, E. *Met. Constructia Massini* **12**, 1018 (1960).
6. AGRINIER, H. *Compt. Rend.* **253**, 280 (1961).
7. ELBEIH, I. I. M. and GABRA, G. G. *Chemist-Analyst* **52**, 77 (1963).
8. HILLEBRAND, W. F., LUNDELL, G. E. F., BRIGHT, H. A. and HOFFMAN, J. I. *Applied Inorganic Analysis* J. Wiley, New York (1954).
9. LADENBAUER, I. M. and SLAMA, O. *Mikrochim. Acta* 903 (1955).
10. CLULEY, H. J. *Analyst* **76**, 523 (1951).
11. HUNT, E. C., NORTH, A. A. and WELLS, R. A. *Analyst* **80**, 172 (1955).
11a. HUNT, E. C. and WELLS, R. A. *Analyst* **79**, 351 (1954).
12. WEATHERLEY, E. G. *Analyst* **81**, 404 (1956).
13. AGRINIER, H. *Compt. Rend.* **257**, 145 (1963).
14. PURUSHOTTAM, B. *Z. Anal. Chem.* **188**, 425 (1962).
15. ELBEIH, I. I. M. and ABOU-ELNAGA, M. A. *Anal. Chim. Acta* **19**, 123 (1958).
16. ACKERMANN, G. and SCHLOSSER, G. *Z. Chem.* **3**, 471 (1963).
17. VENTURELLO, G. and GHE, A. M. *Ann. Chim. (Roma)* **44**, 960 (1954).
18. SCOTT, I. A. P. and MAGEE, R. J. *Talanta* **1**, 329 (1958).
19. DE ANDRADE DIAS, B. and PINTO COELHO, F. *Rev. Fac. A. Univ. Coimbra* **26**, 64 (1957).
20. VAECK, S. V. *Anal. Chim. Acta* **12**, 443 (1955).
21. LIANG Y.-CH. *Hua Hsueh Pao* **24**, 205 (1958).
22. VENTURELLO, G. and GHE, A. M. *Anal. Chim. Acta* **7**, 268 (1952).
23. MUNSHI, K. N. and DEY, A. K. *J. Prakt. Chem.* **18**, 233 (1962).
24. ŠTEFFEK, M. *Chem. Listy* **55**, 1221 (1961).
24a. ŠTEFFEK, M. *Chem. Listy* **56**, 951 (1962).
24b. ŠTEFFEK, M. *Chem. Listy* **57**, 972 (1963).
24c. ŠTEFFEK, M. *Chem. Listy* **58**, 957 (1964).
24d. ŠTEFFEK, M. *Chem. Listy* **61**, 654 (1967).
25. QUENTIN, K. E. *Z. Anal. Chem.* **140**, 92 (1953).
26. HERMANOWICZ, W. and SIKOROWSKA, C. *Gas. Woda. Techn. Sanit.* **29**, 360 (1955).
27. HERMANOWICZ, W. and SIKOROWSKA, C. *Przemysl Chem.* **8**, 238 (1952).
28. GANTCHEV, N. and KOEV, K. *Zh. Analit. Khim.* **17**, 166 (1962).
29. KAPLAN, B. YA. and OLSHEVSKAYA, I. V. *Zavodsk. Lab.* **29**, 26 (1963).
30. BOVALINI, E. and PIAZZI, M. *Ann. Chim. (Roma)* **49**, 1075 (1959).
31. GAUTHIER, P. *Bull. Soc. Chim. France* 981 (1955).
32. BURSTALL, F. H. and WELLS, R. A. *Analyst* **76**, 396 (1951).
33. RYAN, W. and WILLIAMS, A. F. *Analyst* **77**, 293 (1952).
34. KEMBER, N. F. *Analyst* **77**, 78 (1952).
35. WILLIAMS, A. F. *Analyst* **77**, 297 (1952).
36. KINGSBURY, G. W. J. and TEMPLE, R. B. F. *Analyst* **77**, 307 (1952).
37. WILLIAMS, A. F. *J. Chem. Soc.* 3155 (1952).
38. BURSTALL, F. H. and WILLIAMS, A. F. *Analyst* **77**, 983 (1952).
39. MERCER, R. A. and WILLIAMS, A. F. *J. Chem. Soc.* 3399 (1952).
40. MERCER, R. A. and WELLS, R. A. *Analyst* **79**, 339 (1954).
41. VENTURELLO, G. and GHE, A. M. *Analyst* **82**, 343 (1957).
42. BISHOP, J. R. *Analyst* **81**, 291 (1956).
43. KEMBER, N. F. and WELLS, R. A. *Analyst* **76**, 579 (1951).
44. CHEW, B. and LINDLEY, G. *Metallurgia* **53**, 45 (1956).
45. ŘEZÁČ, Z. and ROUBAL, M. *Chem. Listy* **51**, 884 (1957); *Collection Czech. Chem. Commun* **23**, 426 (1958).
46. ŠULCEK, Z., MICHAL, J. and DOLEŽAL, J. *Collection Czech. Chem. Commun.* **24**, 1815 (1959).
47. ŠULCEK, Z., MICHAL, J. and DOLEŽAL, J. *Chemist-Analyst* **50**, 13 (1961).
48. ŠULCEK, Z., MICHAL, J. and DOLEŽAL, J. *Collection Czech. Chem. Commun.* **26**, 246 (1961).
49. ŠULCEK, DOLEŽAL, J., MICHAL, J. and SYCHRA, V. *Talanta* **10**, 3 (1963).
50. HANSEN, R. and GUNNAR, K. *J. Am. Chem. Soc.* **71**, 4158 (1949).

51. VYDRA, F. *Talanta* **11**, 433 (1964).
52. VYDRA, F. and MARKOVÁ, V. *Collection Czech. Chem. Commun.* **30**, 2382 (1965).
53. DOLEŽAL, J. *private communication.*
54. BOZHEVOLNOV, E. A. and KARAKOVSKAYA, O. A. *Zavodsk. Lab.* **27**, 11 (1961).
55. VYDRA, F. and MARKOVÁ, V. *Talanta* **10**, 339 (1963).
56. VYDRA, F. *Talanta* **10**, 753 (1963).
57. KOHLSCHÜTTER, H. W., GETTROST, H., HOFFMANN, G. and STAMM, H. H. *Z. Anal. Chem.* **166**, 262 (1959).
58. KOHLSCHÜTTER, H. W. and GETTROST, H. *Z. Anal. Chem.* **167**, 264 (1959).
59. KUTEINIKOV, A. F. and BRODSKAYA, V. M. *Zh. Analit. Khim.* **17**, 305 (1962).
60. RYDBERG, J. *Acta Chem. Scand.* **9**, 1252 (1955).
61. AHRLAND, S., GRENTHE, J. and NOREN, B. *Acta Chem. Scand.* **14**, 1077 (1960).
62. MARJANOVIĆ-KRAJOVAN, V., TURINA, S., INDJIĆ, N. and JOVANOVIĆ, V. *Z. Anal. Chem.* **195**, 427 (1963).
63. ACKERMANN, G. and SCHLOSSER, G. *Z. Chemie. Leipzig* **5**, 66 (1965).
64. GEL'MAN, E. M. *Zh. Analit. Khim.* **23**, 736 (1968).
65. VIKTOROVA, M. E. and SALTYKOVA, V. S. *Zh. Analit. Khim.*, **23**, 55 (1968).
66. PLAMONDON, J. *Econ. Geol.* **63**, 76 (1968).
67. FUKASAWA, T., FUJII, T. and MIZUIKO, A. *Bunseki Kagaku* **17**, 713 (1968).
68. SUPIŃSKI, J., *Chem. Analit.* (*Warsaw*) **13**, 833 (1968).
69. VYDRA, F. and STARÁ, V. *Collection Czech. Chem. Commun.* **33**, 3883 (1968).

APPENDIX

A review of some separations of various combinations of cations in various solvent systems

The table lists published solvent systems and R_F values of ions chromatographed in a given system. The least difference in R_F values between two separated ions is 0.1 under suitably chosen conditions. The R_F values should not be considered as true constants, but only as informative values; they are reproducible only if the original conditions of separation are strictly observed. The data in the table should serve primarily to give quick information regarding the mobility of the listed ions in the respective solvent systems, and also for facilitating the choice of the most suitable solvent systems for the separation of mixtures. When an R_F value is not given, the ion concerned was not investigated in the corresponding solvent system.

C = comet shaped spot; + = second front; * = two spots. Roman numerals indicate oxidation state. The reference numbers refer to the references in Chap. 6 (pp. 138—151).

Solvent system

Element	MeOH 99%	MeOH + EtOH (1:1)	MeOH + AmOH (7:3)	MeOH + 5M HCl (3:2)	MeOH + AcOH (98:2)	MeOH + 15M NH₃ (95:5)	MeOH + PhOH + HCl (3:5:2)
Ag	—	—	—	—	—	—	—
Al	—	—	—	—	—	—	0.11
Au	—	—	—	—	—	—	—
As	—	—	—	—	—	—	—
Ba	—	0.28	—	—	0.39	—	—
Be	—	—	—	—	0.88	—	—
Bi	—	—	—	—	—	—	—
Ca	—	0.62	—	0.60	0.66	—	—
Cd	—	—	—	—	—	—	—
Co	—	—	—	—	—	—	—
Cr	—	—	—	—	—	—	—
Cs	—	—	—	—	—	0.10	—
Cu	—	—	—	—	—	—	—
Fe	—	—	—	—	—	—	—
Ga	—	—	—	—	—	—	0.78
Ge	—	—	—	—	—	—	—
Hg	—	—	—	—	—	—	—
Hf	—	—	—	—	—	—	—
In	—	—	—	—	—	—	0.30
K	0.49	—	0.09	0.23	—	0.12	—
Li	0.88	—	0.63	—	—	0.74	—
Mg	—	0.80	—	0.79	0.74	—	—
Mn	—	—	—	—	—	—	—
Mo	—	—	—	—	—	—	—
Na	0.72	—	0.25	0.40	—	0.30	—
Nb	—	—	—	—	—	—	—
NH₄	—	—	0.33	—	—	—	—
Ni	—	—	—	—	—	—	—
Pb	—	—	—	—	—	—	—
Rb	—	—	—	—	—	0.12	—
Sb	—	—	—	—	—	—	—
Se	—	—	—	—	—	—	—
Sn	—	—	—	—	—	—	—
Sr	—	0.49	—	—	0.52	—	—
Ta	—	—	—	—	—	—	—
Te	—	—	—	—	—	—	—
Ti	—	—	—	—	—	—	—
Tl	—	—	—	—	—	—	0.67
Th	—	—	—	—	—	—	—
U	—	—	—	—	—	—	—
V	—	—	—	—	—	—	—
W	—	—	—	—	—	—	—
Zn	—	—	—	—	—	—	—
Zr	—	—	—	—	—	—	—
Ref.	[217]	[208]	[220]	[207]	[213]	[225]	

Solvent system

Element	MeOH + EtOH + 2M HCl (3:3:4)	MeOH + 25% H_2SO_4 + H_2O (7:1:4)	MeOH + H_2O + HCl (8:1:1)	MeOH + BuOH + COL + 6M AcOH (2:1:1:1)	MeOH + HCl + THF (2:7:1)	MeOH + H_2O + NH_3 (55:15:30)	2-ethylhexanol + MeOH (3:7)
Ag	0.13	—	—	—	—	—	—
Al	—	—	—	0.88	—	—	—
Au	0.91	—	—	—	—	—	—
As	(III) 0.92	0.00	—	—	—	—	—
Ba	—	—	0.11	0.09	0.17	—	—
Be	—	—	0.85	—	—	—	—
Bi	0.88	—	—	—	—	—	—
Ca	—	—	0.47	0.47	0.43	—	—
Cd	0.96	—	—	—	—	0.75	—
Co	0.70	—	—	0.81	—	0.15	—
Cr	0.73	—	—	0.88	—	—	—
Cs	—	—	0.12	—	—	—	—
Cu	0.65	0.82	0.66	0.83	—	0.42	—
Fe	0.73	—	—	(III) 0.91	—	—	—
Ga	—	—	—	—	—	—	—
Ge	—	—	—	—	—	—	—
Hg	(II) 1.0	0.71	—	(II) 0.88	—	—	—
Hf	—	—	—	—	—	—	—
In	—	—	—	—	—	—	—
K	—	—	0.12	0.08	—	—	0.06
Li	—	—	0.74	0.52	—	—	0.65
Mg	—	—	0.70	0.59	0.59	—	—
Mn	0.70	—	—	0.66	—	—	—
Mo	0.60	—	—	—	—	—	—
Na	—	—	0.31	0.22	—	—	0.21
Nb	0.86	—	—	—	—	—	—
NH$_4$	—	—	0.43	—	—	—	—
Ni	0.63	—	0.69	0.83	—	0.62	—
Pb	0.45	—	—	0.55	—	—	—
Rb	—	—	0.12	—	—	—	—
Sb	0.80	—	—	—	—	—	—
Se	0.72	—	—	—	—	—	—
Sn	(II) 0.94	—	—	—	—	—	—
Sr	—	—	0.26	0.22	0.32	—	—
Ta	1.00	—	—	—	—	—	—
Te	0.71	—	—	—	—	—	—
Ti	0.76	—	—	—	—	—	—
Tl	—	0.35	—	—	—	—	—
Th	—	—	—	—	—	—	—
U	0.76	—	—	—	—	—	—
V	0.74	—	—	—	—	—	—
W	1.00	—	—	—	—	—	—
Zn	0.94	0.84	—	0.89	—	0.44	—
Zr	—	—	—	—	—	—	—
Ref.	[160]	[109]	[141, 184]	[150]	[211]	[164]	[216]

Solvent system

Element	MeOH + PhOH + HCl (1:1:1)	MeOH + PhOH + HCl (45:115:40)	MeOH + PY + DX (7:2:1)	MeOH + iPrOH + AcOH + H_2O (5:3:1:1)	MeOH + DX + cyclohexane (6:10:84)	MeOH + EtOH + 11% NH_4SCN (5:5:2)
Ag	—	—	—	—	—	0.5
Al	0.18	0.10	—	—	—	—
Au	—	—	—	—	—	—
As	—	—	—	—	—	—
Ba	—	0.00	—	—	—	0.25
Be	—	0.23	—	—	—	—
Bi	—	0.20	—	—	—	—
Ca	—	0.70	—	—	—	—
Cd	—	0.30	—	—	—	—
Co	—	0.16	—	0.58	0.64	—
Cr	—	0.12	—	—	—	—
Cs	—	0.37	—	—	—	0.24
Cu	—	0.22	—	—	0.27	—
Fe	—	0.39	—	—	—	—
Ga	0.66	0.66	—	—	—	—
Ge	—	—	—	—	—	—
Hg	—	(II) 0.34	—	—	—	—
Hf	—	—	—	—	—	—
In	0.26	0.20	—	—	—	—
K	—	0.11	—	—	—	—
Li	—	0.30	—	—	—	—
Mg	—	0.17	—	—	—	—
Mn	—	0.16	—	0.34	—	—
Mo	—	—	—	—	—	—
Na	—	0.11	—	—	—	—
Nb	—	—	—	—	—	0.10
NH_4	—	0.25	—	—	—	—
Ni	—	0.14	—	0.66	0.00	—
Pb	—	—	—	—	—	—
Rb	—	0.17	—	—	—	—
Sb	—	(III) 0.44	—	—	—	0.00
Se	—	—	—	—	—	—
Sn	—	—	—	—	—	—
Sr	—	0.03	—	—	—	0.36
Ta	—	—	—	—	—	—
Te	—	—	—	—	—	0.00
Ti	—	(IV) 0.37	—	—	—	—
Tl	0.43	0.53	—	—	—	—
Th	—	—	0.13	—	—	—
U	—	—	0.93	—	—	0.00
V	—	—	—	—	—	—
W	—	—	—	—	—	—
Zn	—	0.35	—	0.95	—	—
Zr	—	—	—	—	—	0.00
Ref.	[287]	[241]	[311]	[329]	[281]	[157]

Solvent system

Element	MeOH + iPrOH + HCOOH (50:30:2) + 2 g ammonium formate	MeOH + EtOH + $1.94M$ HNO_3 (7:3:10)	EtOH 99%	EtOH + Me_2CO (9:1) (PY complexes)	EtOH + H_2O + propionic acid + NH_3 (100:10:5:5)	EtOH + H_2O + propionic acid (100:10:1)
Ag	—	—	—	—	—	—
Al	—	—	—	—	—	—
Au	—	—	—	—	—	—
As	—	—	—	—	—	—
Ba	0.30	—	—	—	0.23	—
Be	—	—	—	—	—	—
Bi	—	—	—	—	—	—
Ca	0.60	—	—	—	0.82	—
Cd	—	—	—	0.48	—	—
Co	—	—	—	0.52	—	—
Cr	—	—	—	—	—	—
Cs	—	—	—	—	—	—
Cu	—	—	—	0.30	—	—
Fe	—	—	—	—	—	—
Ga	—	—	—	—	—	—
Ge	—	—	—	—	—	—
Hg	—	—	—	—	—	—
Hf	—	—	—	—	—	—
In	—	—	—	—	—	—
K	—	—	0.16	—	—	0.17
Li	—	—	0.70	—	—	0.68
Mg	0.75	—	—	—	0.93	—
Mn	—	—	—	—	—	—
Mo	—	—	—	—	—	—
Na	—	—	0.32	—	—	0.31
Nb	—	—	—	—	—	—
NH_4	—	—	—	—	—	—
Ni	—	—	—	0.50	—	—
Pb	—	—	—	—	—	—
Rb	—	—	—	—	—	—
Sb	—	—	—	—	—	—
Se	—	(VI) (IV) 0.92 0.79	—	—	—	—
Sn	—	—	—	—	—	—
Sr	0.45	—	—	—	0.56	—
Ta	—	—	—	—	—	—
Te	—	(VI) (IV) 0.00 0.67	—	—	—	—
Ti	—	—	—	—	—	—
Tl	—	—	—	—	—	—
Th	—	—	—	—	—	—
U	—	—	—	—	—	—
V	—	—	—	—	—	—
W	—	—	—	—	—	—
Zn	—	—	—	0.70	—	—
Zr	—	—	—	—	—	—
Ref.	[206, 210]	[260]	[217]	[157]	[209]	[224]

Solvent system

Element	EtOH + HCl (6:1)	EtOH + HCl (4:1)	EtOH + 5M HCl (9:1)	EtOH + 8M HNO₃ (9:1)	EtOH + 1M HNO₃ (1:1)	EtOH + 4M H₂SO₄ (4:1)
Ag	—	—	0.02	0.18	0,60	—
Al	—	0.61	0.37	0.41	—	—
Au	—	—	0.95	—	—	—
As	—	—	0.50	—	—	—
Ba	—	—	0.04	—	—	—
Be	—	—	0.70	—	—	—
Bi	—	—	0.94	0.67	—	—
Ca	—	—	—	—	—	—
Cd	—	—	1.00	0.37	—	—
Co	—	0.52	0.32	0.32	—	—
Cr	—	0.56	0.47	(III) (VI) 0.43 C	—	—
Cs	—	—	—	—	—	—
Cu	—	0.61	0.47	0.37	0.84	0.41
Fe	—	0,79	0.56	(III) 0.39	—	0.57
Ga	—	—	—	—	—	—
Ge	—	—	—	0.54	—	—
Hg	—	—	(I) (II) 0.08C 1.0	0.70C	0.70	—
Hf	—	—	—	—		
In	—	—	—	—		
K	—	—	0.08	—		
Li	—	—	—	—		
Mg	—	—	0.33	—	—	—
Mn	—	0.60	0.36	(II) 0.38	—	—
Mo	—	0.71	0.37	0.45	—	—
Na	—	—	—	—	—	—
Nb	—	—	—	—		
NH₄	—	—	—	—		
Ni	—	0.52	0.34	0.26	—	—
Pb	—	—	0.16	0.11	—	—
Rb	—	—	—	—		
Sb	—	—	0.85	—	—	—
Se	0.72	—	—	—	—	—
Sn	—	0.95	0.97	—	—	—
Sr	—	—	0.11	—	—	—
Ta	—	—	—	—		
Te	0.94	—	—	—		
Ti	—	0.51	0.50	C	—	—
Tl	—	—	(I) (III) 0.001.00	—	—	—
Th	—	0.14	0.12	—	—	—
U	—	0.57	0.57	—	—	0.72
V	—	0.54	0.38	0.49	—	—
W	—	—	—	—	—	—
Zn	—	—	0.93	0.32	—	—
Zr	—	0.03	—	—	—	—
Ref.	[263]	[265]	[241]	[275]	[84]	[240]

Solvent system

Element	EtOH + 2M AcOH (4:1)	EtOH + 7M AcOH (9:1)	EtOH + H$_2$O (4:1)	EtOH + H$_2$O (87:13)	EtOH + HCl + H$_2$O (15:4:1)	EtOH + HCl + H$_2$O (5:1:4)
Ag	—	—	0.5	—	—	—
Al	—	—	—	—	0.68	—
Au	—	—	—	—	—	—
As	—	—	—	—	—	—
Ba	0.43	—	—	0.14	—	—
Be	—	—	—	—	—	—
Bi	—	—	—	—	0.62	—
Ca	0.68	0.47	—	0.42	0.48	—
Cd	—	—	—	—	0.65	—
Co	—	—	—	—	0.59	0.64
Cr	—	—	—	—	0.62	—
Cs	—	—	—	0.18	—	—
Cu	—	—	—	—	0.68	0.35
Fe	—	—	—	—	0.74	—
Ga	—	—	—	—	—	—
Ge	—	—	—	—	—	—
Hg	—	—	<0.5	—	—	—
Hf	—	—	—	—	—	—
In	—	—	—	—	—	—
K	0.45	0.20	—	0.18	—	—
Li	0.76	—	—	0.71	—	—
Mg	0.76	0.63	—	0.59	0,60	—
Mn	—	—	—	—	0.62	—
Mo	—	—	—	—	0.71	—
Na	0.56	0.30	—	0.32	—	—
Nb	—	—	—	—	—	—
NH$_4$	—	—	—	0.50	—	—
Ni	—	—	—	—	0.57	0.58
Pb	—	—	<0.5	—	—	—
Rb	—	—	—	0.18	—	—
Sb	—	—	—	—	—	—
Se	—	—	—	—	—	—
Sn	—	—	—	—	0.96	—
Sr	0.55	—	—	0.24	—	—
Ta	—	—	—	—	—	—
Te	—	—	—	—	—	—
Ti	—	—	—	—	0.60	—
Tl	—	—	—	—	—	—
Th	—	—	—	—	0.34	—
U	—	—	—	—	0.70	—
V	—	—	—	—	0.62	—
W	—	—	—	—	—	—
Zn	—	—	—	—	0.57	—
Zr	—	—	—	—	0.09	—
Ref.	[145]	[207]	[88]	[204]	[265]	[242]

Solvent system

Element	EtOH + iPrOH + 5M HCl (9:9:2)	EtOH + Me$_2$CO NaBr + NH$_4$OAc (90:10:4:1) (PY complexes)	EtOH + MeOH + HCl + H$_2$O (10:6:8:3)		nPrOH + 7M LiNO$_3$	nPrOH + H$_2$O + EtOAc + HCl + HNO$_3$ (30:5:40:4:15)	iPrOH + 5M HCl (9:1)		iPrOH + 8M HNO$_3$ (9:1)
Ag	0.02	—	—		—	—	0.06		0.13
Al	0.23	—	—		—	—	0.35		0.05
Au	0.93	—	—		—	1.00	1.00		—
As	0.43	—	(III) 0.55	(V) 0.75	—	—	0.66		—
Ba	0.02	—	—		0—0.33	—	0.05		—
Be	—	—	—		—	—	—		—
Bi	0.68	—	—		—	—	0.84		0.03C
Ca	—	—	—		0.65	—	—		—
Cd	0.75	0.48	—		—	—	0.84		0.07
Co	0.15	0.52	—		—	0.15	0.27		0.03
Cr	0.17	—	(III) 0.3	(VI) 0.8	—	—	0.28		(III) 0.05
Cs	—	—	—		—	—	—		—
Cu	0.27	0.30	(I) 0.90	(II) 0.35	—	0.40	0.28		0.05
Fe	0.42	—	—		—	0.70	0.35		(III) 0.06
Ga	—	—	—		—	—	—		—
Ge	—	—	—		—	—	—		0.25
Hg	—	—	—		—	—	(I) 0.05	(II) 1.00	0.58
Hf	—	—	—		—	—	—		—
In	—	—	—		—	—	—		—
K	—	—	—		—	—	0.15		—
Li	—	—	—		—	—	—		—
Mg	—	—	—		—	—	0.23		—
Mn	0.22	—	—		—	—	0.37		—
Mo	0.20	—	(V) 0.20	(VI) 0.55	—	—	0.28		0.20
Na	—	—	—		—	—	—		—
Nb	—	—	—		—	—	—		—
NH$_4$	—	—	—		—	—	—		—
Ni	0.15	—	—		—	0.15	0.23		0.02
Pb	0.04	—	—		—	—	0.03		0.02
Rb	—	—	—		—	—	—		—
Sb	—	—	(III) 0.75	(V) 0.00	—	1.00	0.77		—
Se	—	—	—		—	—	—		—
Sn	0.86	—	—		—	—	0.88		—
Sr	0.07	—	—		0.49	—	0.11		—
Ta	—	—	—		—	—	—		—
Te	—	—	—		—	—	—		—
Ti	0.32	—	—		—	—	0.33		0.14C
Tl	0.98	—	—		—	—	1.00		—
Th	—	—	—		—	—	—		—
U	0.29	—	—		—	—	0.36		—
V	—	—	—		—	—	—		0.27
W	—	—	—		—	—	—		—
Zn	0.88	0.70	—		—	—	0.87		0.05
Zr	—	—	—		—	—	—		—
Ref.	[241]	[157]	[110]		[183]	[121]	[241]		[275]

Solvent system

Element	iPrOH + AcOH + H_2O (8:1:1)	iPrOH + AcMe$_2$CO + HCl (3:1:1)	iPrOH + PY + H_2O + AcOH (8:8:4:1)	AmOH satd. with $2M$ HCl	AmOH + BuOAc + MeOH + H_2O + HCl (20:30:3:2:15)	AllOH + H_2O + EtOAc + AcOH (5:3:3:3)
Ag	0.36	—	—	0.00	—	0.75
Al	—	—	0.69	—	—	—
Au	—	—	—	0.36	1,00	—
As	—	—	—	0.41	0.82	—
Ba	—	—	0.09	—	—	—
Be	—	0.60	0.91	—	—	—
Bi	0.51	—	—	0.22	—	0.74
Ca	—	—	0.35	—	—	—
Cd	0.39	—	0.97	0.23	—	—
Co	—	—	0.92	0.03	0.52	—
Cr	—	—	—	—	0.27	—
Cs	—	—	0.16	—	—	—
Cu	0.44	—	0.88	0.02	0.60	0.62
Fe	—	—	0.96	0.02	1.00	0.94
Ga	—	—	—	—	—	—
Ge	—	—	—	—	—	—
Hg	0.69	—	—	(II) 0.55	—	(II) 0.87
Hf	—	—	—	—	—	—
In	—	—	—	—	—	—
K	—	—	0.13	—	—	—
Li	—	—	0.61	—	—	—
Mg	—	—	0.53	—	—	—
Mn	—	—	0.84	—	—	—
Mo	—	—	—	—	—	—
Na	—	—	0.24	—	—	—
Nb	—	—	—	—	—	—
NH$_4$	—	—	—	—	—	—
Ni	—	—	0.92	0.00	0.25	—
Pb	0.24	—	0.71	0.00	0.43	0.48
Rb	—	—	0.15	—	—	—
Sb	—	—	—	—	0.98	—
Se	—	—	—	—	—	—
Sn	—	—	—	0.52	—	—
Sr	—	—	0.19	—	—	—
Ta	—	—	—	—	—	—
Te	—	—	—	—	—	—
Ti	—	1.00	—	0.02	—	—
Tl	—	—	—	0.75	—	—
Th	—	—	—	—	—	—
U	—	0.83	—	—	0.70	—
V	—	—	—	—	0.38	—
W	—	—	—	—	—	—
Zn	—	—	0.91	—	—	—
Zr	—	—	—	—	—	—
Ref.	[86]	[149]	[143]	[241]	[158]	[158]

Solvent system

Element	BuOH satd. with 1M HCl	BuOH satd. with 3M HCl	BuOH satd. with 6M HCl	BuOH satd. with 10% HBr	BuOH satd. with 10% HNO_3	BuOH + HCl + H_2O (114:31:55)	BuOH+HCl+H_2O (407:400:193)	BuOH + HCl + 3% H_2O_2 (1000:47:200)
Ag	0.00	0.00	—	—	0.12	—	—	—
Al	0.04	0.07	—	$<$0.1	$<$0.07	—	—	—
Au	1.05.1.13+	—	—	—	—	—	—	—
As	(III) (V) 0.52 0.84	0.83	—	(III) 0.51	—	0.86	—	—
Ba	—	—	—	—	—	—	—	—
Be	0.15	0.30	0.58	—	—	—	—	—
Bi	0.65	0.61	—	0.71	0.25	—	0.55	—
Ca	—	0.05	—	—	—	—	—	—
Cd	0.60	0.90	—	0.69	—	—	0.51	—
Co	—	0.05	—	$<$0.10	$<$0.07	0.30	—	—
Cr	—	0.05	—	—	$<$0.07	—	0.28	—
Cs	—	—	—	—	—	—	—	—
Cu	0.10	0.26	—	0.09	0.08	0.36	—	—
Fe	0.15	0.50—0.60	—	$<$0.10	(III) $<$0.07	(III) (II) 0.45 0.75	0.94	—
Ga	0.44	0.42	1.00	—	—	—	0.97	—
Ge	0.26	0.54	0.93	0.28	0.23	—	0.98	—
Hg	0.96	0.81	—	(II) 0.86	C	0.98	0.74	—
Hf	—	—	—	—	—	—	—	—
In	0.24	0.40	0.55	—	—	—	0.44	—
K	—	—	—	—	—	—	—	—
Li	—	—	—	—	—	—	—	—
Mg	—	0—0.10	—	—	—	—	—	—
Mn	—	0.05	—	$<$(II) 0.1	$<$0.07	0.30	0.31	—
Mo	—	0.45—0.55	—	—	0.33	—	—	—
Na	—	0.05	—	—	—	—	—	—
Nb	—	—	—	—	—	—	—	—
NH_4	—	—	—	—	—	—	—	—
Ni	0.01	0.05	—	0.26	$<$0.07	0.32	—	—
Pb	0.00	0.05	—	0.11	0.06	0.37	0.38	—
Rb	C	—	—	—	—	—	—	—
Sb	—	0.34	—	—	—	—	0.68	0.00
Se	—	—	—	—	—	—	0.88	—
Sn	(II) 0.97	0.99	—	—	—	—	0.74	(IV) 0.7
Sr	—	—	—	—	—	—	—	—
Ta	—	—	—	—	—	—	—	—
Te	—	(IV) (VI) 0.56 0.11	—	—	—	—	0.83	—
Ti	—	0—0.10	—	C	—	—	—	—
Tl	—	—	—	—	—	—	0.97	—
Th	—	0.05	—	—	—	—	0.15	—
U	—	0.26	—	—	—	—	—	—
V	—	—	—	0.49	0.20	—	0.38	—
W	—	—	—	—	—	—	—	—
Zn	0.71	—	—	0.32	$<$0.07	—	0.74	—
Zr	—	0.05	—	—	—	—	—	—
Ref.	[241]	[112]	[149]	[275]	[275]	[102]	[169]	[123]

Solvent system

Element	BuOH + HCl + 30% H_2O_2 (10:1:1:8)	BuOH + AcOH + EtOAc (10:2:1)	BuOH + $PhCH_2OH$ + 0.1M HNO_3 (100:1:100)	BuOH + 1M HNO_3 + antipyrine (100:100:1)	BuOH + 20M HF + HCl + H_2O (50:1:25:24)	BuOH satd. with $PhCH_2OH$ + 3M HCl + Me_2CO + iPrOH (6:10:3:2)
Ag	—	—	0.24	0.19	—	—
Al	—	—	—	—	—	—
Au	—	—	—	—	—	—
As	—	—	—	0.49	—	—
Ba	—	—	—	—	—	—
Be	—	—	—	—	—	—
Bi	—	—	—	0.20	0.68	—
Ca	—	—	—	—	—	—
Cd	—	—	—	0.12	—	—
Co	—	—	—	—	—	0.08
Cr	—	(II) 0.23 (III) 0.75 (VI) 0.10	—	—	—	—
Cs	—	—	—	—	—	—
Cu	—	—	—	0.17	—	0.16
Fe	—	—	—	—	0.48	(II) 0.68
Ga	—	—	—	—	—	—
Ge	—	—	—	—	—	—
Hg	—	—	—	—	—	—
Hf	—	—	—	—	0.50	—
In	—	—	—	—	—	—
K	—	—	—	—	—	—
Li	—	—	—	—	—	—
Mg	—	—	—	—	—	—
Mn	—	—	—	—	—	—
Mo	0.79	—	—	—	—	—
Na	—	—	—	—	—	—
Nb	—	—	—	—	0.64	—
NH_4	—	—	—	—	—	—
Ni	—	—	—	—	—	0.08
Pb	—	—	0.16	0.10	—	0.29
Rb	—	—	—	—	—	—
Sb	—	—	—	—	—	—
Se	—	—	—	—	—	—
Sn	—	—	—	(IV) 0.67 (II) 0.72	—	—
Sr	—	—	—	—	—	—
Ta	—	—	—	—	1,00	—
Te	—	—	—	—	—	—
Ti	0.09	—	—	—	0.57	—
Tl	—	—	0.18	—	—	—
Th	—	—	—	—	—	—
U	—	—	—	—	0.39	—
V	0.27	—	—	—	—	—
W	—	—	—	—	—	—
Zn	—	—	—	—	—	0.69
Zr	—	—	—	—	0.50	—
Ref.	[305]	[297]	[94]	[95]	[126]	[156]

Solvent system

Element	BuOH + NH$_3$ + PY (8:1:2)	BuOH + iPrOH + 5M HCl (9:9:2)	BuOH + PrOH (1:1) satd. with 7M LiNO$_3$/2M HNO$_3$	BuOH + AcOH + satd. ROAc + H$_2$O (10:2:1:7)	BuOH + AcMe$_2$CO + 2M HNO$_3$ (50:1:50)
Ag	—	0.03	—	0.12	0.15
Al	—	0.08	—	0.15	0.09
Au	—	0.89	—	—	—
As	—	0.44	—	0.15	0.43
Ba	—	0.04	—	0.17	0.09
Be	—	—	—	—	—
Bi	—	0.65	—	0.34	0.23
Ca	—	—	—	0.20	0.08
Cd	0.51	0.73	—	0.28	0.12
Co	—	0.08	—	0.24	0.10
Cr	—	0.07	—	0.65	0.09
Cs	—	—	—	—	—
Cu	—	0.12	—	0.63	0.12
Fe	—	0.11	—	0.73	0.43
Ga	—	—	—	0.24	—
Ge	—	—	—	—	—
Hg	—	—	—	(I) 0.09 (II) 0.83	0.43
Hf	—	—	—	—	—
In	—	—	—	—	—
K	—	—	—	0.23	0.10
Li	—	—	—	—	—
Mg	—	—	—	0.20	0.10
Mn	—	0.08	—	0.25	0.11
Mo	—	0.23	—	—	—
Na	—	—	—	0.25	0.10
Nb	—	—	—	—	—
NH$_4$	—	—	—	—	—
Ni	—	0.09	—	0.24	0.09
Pb	—	0.02	—	0.18	0.09
Rb	—	—	—	—	—
Sb	—	—	—	0.13	0.02
Se	—	—	0.72	—	—
Sn	—	0.82	—	0.14	0.82
Sr	—	0.05	—	0.18	0.07
Ta	—	—	—	—	—
Te	—	—	0.47	—	—
Ti	—	0.04	—	0.07	—
Tl	—	0.94	—	—	—
Th	—	—	—	—	—
U	—	0.14	—	—	—
V	—	—	—	—	—
W	—	—	—	—	—
Zn	0.29	0.80	—	0.30	0.10
Zr	—	—	—	—	—
Ref.	[161]	[241]	[262]	[281]	[212]

Solvent system

Element	BuOH satd. with HBr + H_2O (1:9) 100 ml org. fraction+10ml Br_2	BuOH satd. with 1M citric acid	iBuOH + HCl (3:1)	iBuOH + 4M HCl	iBuOH + H_2O + PY (85:10:5) 2.1M in HCNS	iBuOH+Me_2CO+ 6M HCl + $AcMe_2CO$ (30:40:2:1)
Ag	0.10C	0.14	—	—	—	—
Al	0.16	—	0.00	—	—	—
Au	0.97	—	—	—	—	—
As	(III) 0.78	(III) 0.43	—	—	—	—
Ba	0.02	—	—	—	—	—
Be	0.46	—	0.33	—	—	—
Bi	1.00	0.06C	—	—	—	—
Ca	0.08	—	—	—	—	—
Cd	1.00	0.03	—	—	—	—
Co	0.14	—	—	—	—	—
Cr	0.06	—	—	—	—	—
Cs	0.11	—	—	—	—	—
Cu	1.00	0.10	—	—	—	0.50
Fe	0.58	—	—	—	—	0.79
Ga	0.10	—	—	—	—	—
Ge	—	—	—	—	—	—
Hg	(II) 1.0	(I) 0.11 (II) 0.74	—	—	—	—
Hf	—	—	—	—	—	—
In	1.00	—	—	—	—	—
K	0.11	—	—	—	—	—
Li	—	—	—	—	—	—
Mg	0.16	—	—	—	—	—
Mn	0.19	—	—	—	—	—
Mo	—	—	—	—	—	—
Na	—	—	—	—	—	—
Nb	—	—	—	—	—	—
NH_4	—	—	—	—	—	—
Ni	0.13	—	—	—	—	—
Pb	0.60	0.05	—	—	—	0.25
Rb	0.11	—	—	—	—	—
Sb	1.00	(III) 0.57	—	—	—	—
Se	—	—	—	—	—	—
Sn	(II) 1.0	0.64	—	—	—	—
Sr	0.04	—	—	—	—	—
Ta	—	—	—	—	—	—
Te	—	—	—	—	—	—
Ti	0.19	—	—	—	—	—
Tl	(III) 1.00	—	—	—	—	—
Th	0.03	—	—	0.09	0.46	—
U	(IV) 0.02 (VI) 0.21	—	—	0.31	—	—
V	0.20	—	—	—	—	—
W	—	—	—	—	—	—
Zn	1.00	—	—	—	—	0.95
Zr	0.01	—	—	—	0.80	—
Ref	[516]	[125]	[144]	[316]	[261]	[162]

Solvent system

Element	tBuOH + H_2O + HCl (7:2:1)	tBuOH + Me_2CO + H_2O + HNO_3 (10:10:3:2)	tBuOH + $CHCl_3$ + 8M HCl + $AcMe_2CO$ (45:45:8:2)	Me_2CO + H_2O (1:1) separation of sulphates	Me_2CO + 6M HCl (19:1)	Me_2CO + 0.5M HCl (19:1)
Ag	0.00	0.25	—	—	0.00	0.40
Al	—	—	0.05	—	—	—
Au	—	—	—	—	1.00	1.00
As	0.51	—	(III) 0.64	—	—	—
Ba	—	—	—	—	—	—
Be	—	—	—	—	—	—
Bi	0.79	(III) 0.76	—	—	0.93	0.93
Ca	—	—	—	—	—	—
Cd	0.97	0.32	—	—	0.63	—
Co	—	—	—	—	0.33	0.80
Cr	—	—	—	—	—	—
Cs	—	—	—	—	—	—
Cu	0.29	0.40	—	—	0.46	0.60
Fe	—	—	—	—	0.76	(III) (II) 0.70 0.80
Ga	—	—	—	—	—	—
Ge	—	—	—	—	—	—
Hg	0.97	(I) (II) 0.00 0.81	—	—	(I) (II) 0.96 0.80	—
Hf	—	—	—	—	—	—
In	—	—	—	—	—	—
K	—	—	—	0.26	—	—
Li	—	—	—	0.52	—	—
Mg	—	—	—	—	—	—
Mn	—	—	—	—	—	—
Mo	—	—	—	—	—	—
Na	—	—	—	0.35	—	—
Nb	—	—	—	—	—	—
NH_4	—	—	—	—	—	—
Ni	—	—	—	—	0.00	0.62
Pb	0.15	0.50	—	—	0.48	0.16
Rb	—	—	—	—	—	—
Sb	0.96	—	(III) 0.85	—	—	—
Se	—	—	—	—	—	—
Sn	—	—	(II) 1.00	—	—	—
Sr	—	—	—	—	—	—
Ta	—	—	—	—	—	—
Te	—	—	—	—	—	—
Ti	—	—	—	—	—	—
Tl	0.02	—	—	—	(I) (III) 0.11 0.00	0.26
Th	—	—	—	—	—	—
U	—	—	—	—	—	—
V	—	—	—	—	—	—
W	—	—	—	—	—	—
Zn	—	—	—	—	—	—
Zr	—	—	—	—	—	—
Ref.	[90]	[98]	[98]	[219]	[92]	[92]

Solvent system

Element	Me₂CO + HCl + AcOH (90:10:2)	Me₂CO + EtOH + HCl (99:1:1)	Me₂CO + MeOAc (4:1)	Me₂CO + 10% antipyrine (19:1)	Me₂CO₂ + BuOH H₂O + H₂O₂ HCl (80:70:40:10:3)	Me₂CO₂ + iBuOH + 12M HCl (60:38:2)	Me₂CO + 5% thioacetamide (19:1)
Ag	—	—	0.00	0.37	—	—	0.00
Al	0.02	—	—	—	—	—	—
Au	—	—	1.00	1.00	—	—	1.00
As	1.00; 0.55*	—	—	—	—	—	—
Ba	—	—	—	—	—	—	—
Be	—	—	—	—	—	—	—
Bi	0.65	—	1.00	0.98	—	—	0.92
Ca	—	—	—	—	—	—	—
Cd	0.06	—	1.00	0.45	—	—	0.46
Co	0.45	1.00	0.00	0.40	—	0.45	0.40
Cr	0.05	—	—	—	—	—	—
Cs	—	—	—	—	—	—	—
Cu	0.55	—	0.00	0.00	—	—	0.00
Fe	1.00	—	0.00	(II) (III) 0.73 0.00	—	—	0.00
Ga	—	—	—	—	—	—	—
Ge	—	—	—	—	0.50	—	—
Hg	0.80	—	—	0.13	—	—	0.77
Hf	—	—	—	—	—	—	—
In	—	—	—	—	—	—	—
K	—	—	—	—	—	—	—
Li	—	—	—	—	—	—	—
Mg	—	—	—	—	—	—	—
Mn	0.25	—	—	—	—	—	—
Mo	0.75C	—	—	—	—	—	—
Na	—	—	—	—	—	—	—
Nb	—	—	—	—	—	—	—
NH₄	—	—	—	—	—	—	—
Ni	0.03	0.00	0.00	0.00	—	—	0.00
Pb	0.75; 0.45*	—	0.00	0.00	—	—	0.23
Rb	—	—	—	—	—	—	—
Sb	0.70C	—	—	—	—	—	—
Se	—	—	—	—	—	—	—
Sn	0.75	—	—	—	0.75	—	—
Sr	—	—	—	—	—	—	—
Ta	—	—	—	—	—	—	—
Te	—	—	—	—	—	—	—
Ti	0.70; 0.08*	—	—	—	—	—	—
Tl	—	—	0.00	0.00	—	—	0.00
Th	—	—	—	—	—	0.00	—
U	—	—	—	—	—	0.53	—
V	0.10	—	—	—	—	—	—
W	—	—	—	—	—	—	—
Zn	0.75	—	—	—	—	—	—
Zr	—	—	—	—	—	—	—
Ref.	[113]	[342]	[92]	[92]	[99]	[272]	[92]

Solvent system

Element	$Me_2CO + 5\%$ Na_2HPO_4 (9:1)	$Me_2CO + HCl +$ $AcMe_2CO$ (90:10:2)	$Me_2CO +$ $MeiBuCO +$ $HNO_3 + H_2O$ (5:4:1:1)	$Me_2CO +$ $2M\ HCl +$ 10% Hox (8:1:1)	$Me_2CO + H_2O$ $+ HF + HCl$ (90:4:1:5)	$Me_2CO + iBuOH$ $+ 6M\ HCl +$ $AcMe_2CO$ (30:4:2:1)
Ag	0.45	—	—	—	0.59	—
Al	—	0.02	—	—	—	—
Au	1.00	—	—	—	0.97	—
As	—	1.0; 0.55[+]	(III) (V) 0.25 0.57	—	(III) 0.65	—
Ba	—	—	—	—	—	—
Be	—	—	—	—	—	—
Bi	0.96	0.65	0.77	—	0.85	—
Ca	—	—	—	—	—	—
Cd	0.68	0.60	—	—	0.77	—
Co	0.55	—	0.08	—	0.45	—
Cr	—	0.05	—	—	0.06	—
Cs	—	—	—	—	—	—
Cu	0.00	—	0.08	—	0.56	0.50
Fe	(III) (II) 0.00 0.52	—	0.05	0.93	0.86	0.79
Ga	—	—	—	—	—	—
Ge	—	—	0.30	—	—	—
Hg	0.83	0.80	1.00	—	(II) 1.00	—
Hf	—	—	—	—	—	—
In	—	—	—	—	—	—
K	—	—	—	—	—	—
Li	—	—	—	—	—	—
Mg	—	—	—	—	—	—
Mn	—	0.25	—	—	0.25	—
Mo	—	0.75C	—	—	0.83	—
Na	—	—	—	—	—	—
Nb	—	—	—	0.84	0.82	—
NH_4	—	—	—	—	—	—
Ni	0.17	—	0.08	—	0.03	—
Pb	0.35	—	0.08	—	0.48	0.25
Rb	—	—	—	—	—	—
Sb	—	0.70C	—	—	0.86	—
Se	—	—	0.35	—	1.00	—
Sn	—	0.75	—	—	(II) 0.82	—
Sr	—	—	—	—	—	—
Ta	—	—	—	0.68	1.00	—
Te	—	—	0.35	—	0.71	—
Ti	—	0.08; 0.7[+]	—	0.45	(IV) 0.64	—
Tl	0.08	—	—	—	—	—
Th	—	—	1.00	—	—	—
U	—	—	1.00	—	0.64	—
V	—	0.10	—	0.75	0.64	—
W	—	—	—	—	1.00	—
Zn	—	0.75	—	—	0.90	0.95
Zr	—	—	—	—	—	—
Ref	[92]	[113]	[108]	[128]	[160]	[162]

Solvent system

Element	Me$_2$CO + H$_2$O + HCl (85 : 5 : 10)	Me$_2$CO + H$_2$O + HCl (7 : 2 : 1)	Me$_2$CO + HCl + H$_2$O (8 : 1 : 2)	AcMe$_2$CO satd. with water and 0.5M HCl + Me$_2$CO (3 : 1)	Et$_2$CO satd. with 2.2M HF and 2M HNO$_3$
Ag	—	0.91	0.92	—	—
Al	0.75	—	0.21	—	—
Au	—	—	—	—	—
As	—	(III) 0.78	0.78	0.18	—
Ba	—	0.06	—	0.45	—
Be	—	0.60	0.57	—	—
Bi	—	0.98	0.97	—	—
Ca	—	0.21	—	—	—
Cd	—	0.98	0.94	—	—
Co	0.85	—	0.26	—	—
Cr	0.60	—	(III) (VI) 0.25 0.97	—	—
Cs	—	—	—		
Cu	—	0.80	0.65	—	—
Fe	1.00	—	0.98	—	—
Ga	—	—	—	—	—
Ge	—	—	—	—	—
Hg	—	0.98	0.99	—	—
Hf	—	—	—	—	—
In	—	—	—	—	—
K	—	—	—	—	—
Li	—	—	—	—	—
Mg	—	0.30	—	—	—
Mn	0.80	—	0.34	—	—
Mo	0.96	—	0.94	—	—
Na	—	—	—	—	—
Nb	—	—	—	—	0.55
NH$_4$	—	—	—	—	—
Ni	0.69	—	0.16	—	—
Pb	—	0.70	0.73	—	—
Rb	—	—	—	—	—
Sb	—	(III) 0.98	0.98	0.53	—
Se	—	—	—	—	—
Sn	—	—	0.98	1.00	—
Sr	—	0.12	—	—	—
Ta	—	—	—	—	1.0
Te	—	0.02	—	—	—
Ti	0.80	—	0.29	—	0.05
Tl	—	—	0.01	—	—
Th	—	—	0.08	—	—
U	0.94	—	0.98	—	—
V	0.88	—	(V) (IV) 0.36 0.94	—	—
W	—	—	—	—	—
Zn	0.73	—	0.97	—	—
Zr	—	—	0.02	—	—
Ref.	[160]	[90]	[89]	[104, 120]	[131]

Solvent system

Element	MeEtCO + 10M HCl (3:1)	MeEtCO + HCl (85:15)	MeEtCO + HCl (92:8)	60 g 49% HF in 100 ml MeEtCo	MeEtCO + DX + H_2O (5:1:1) 0.25M in Hox and 0.1M in HCl	MeiBuCO + 40% HF (100:3)	THF + HCl (50 + 15)	THF + 7.44M HCl (3:7)	THF + Et_2O (50:15)
Ag	—	—	—	—	—	—	—	—	—
Al	—	0.19	—	—	—	—	0.35	—	—
Au	—	—	—	—	—	—	—	—	—
As	—	—	—	—	—	—	—	—	—
Ba	—	—	—	—	—	—	—	0.28	—
Be	—	0.51	—	—	—	—	—	0.92	—
Bi	—	—	—	0.00	—	—	—	—	—
Ca	—	—	—	—	—	—	0.20	0.67	—
Cd	—	—	—	—	—	—	1.00	—	—
Co	—	—	—	—	—	—	0.78	—	0.78
Cr	—	—	0.54	—	—	—	—	—	—
Cs	—	—	—	—	—	—	—	—	—
Cu	—	—	0.71	—	—	—	0.90	—	0.91
Fe	1.00	1.00	0.93	—	—	—	1.00	—	—
Ga	—	—	—	—	—	—	1.00	—	—
Ge	—	—	—	—	—	—	—	—	—
Hg	—	—	—	—	—	—	—	—	—
Hf	—	—	—	—	—	—	—	—	—
In	—	—	—	—	—	—	—	—	—
K	—	—	—	—	—	—	—	—	—
Li	—	—	—	—	—	—	—	—	—
Mg	—	—	—	—	—	—	0.34	0.84	—
Mn	—	—	0.18	—	—	—	0.50	—	0.51
Mo	—	—	—	—	0.70	—	—	—	—
Na	—	—	—	—	—	—	—	—	—
Nb	0.78	—	—	—	0.40	0.10	—	—	—
NH$_4$	—	—	—	—	—	—	—	—	—
Ni	—	—	0.01	—	—	—	0.28	—	0.27
Pb	—	—	—	—	—	—	—	—	—
Rb	—	—	—	—	—	—	—	—	—
Sb	—	—	—	—	—	—	—	—	—
Se	—	—	—	1.00	—	—	—	—	—
Sn	—	—	—	—	—	—	—	—	—
Sr	—	—	—	—	—	—	—	0.49	—
Ta	0.11	—	—	—	—	0.87	—	—	—
Te	—	—	—	0.63	—	—	—	—	—
Ti	0.10—0.38	—	—	—	—	—	—	—	—
Tl	—	—	—	—	—	—	—	—	—
Th	—	—	—	—	—	—	—	—	—
U	—	—	0.87	—	—	—	—	—	0.97
V	—	—	0.14	—	—	—	—	—	0.55
W	—	—	—	—	—	—	—	—	—
Zn	—	—	—	—	—	—	1.00	—	—
Zr	—	—	—	—	0.10	—	—	—	—
Ref.	[127]	[151]	[120]	[263]	[139]	[129]	[205]	[142]	[304]

Solvent system

Element	THS + HNO$_3$ + H$_2$O (85:10:5)	Et$_2$O + HNO$_3$ (99:1)	Et$_2$O + MeOH + H$_2$O + HNO$_3$ (50:30:20:2)	Light petroleum MeOH + DX (92:7:1)	Light petroleum MeOH + DX (92:7:1)	MeOAc + HNO$_3$ + H$_2$O (85:5:10)
Ag	—	—	0.32	—		—
Al	—	—	—	Separation in the		—
Au	—	—	—	form of complexes		—
As	—	—	—	with thenoyl per-		—
Ba	—	—	—	fluorobutyrylme-		—
				thane		
Be	—	—	—	—	—	—
Bi	—	—	0—0.8C	—	—	—
Ca	—	—	—	—	—	—
Cd	—	—	0.60	—	—	—
Co	—	—	0.61	0.00	0.46	—
Cr	—	—	—	—	—	—
Cs	—	—	—	—	—	—
Cu	—	—	0.61	0.48	0.77	—
Fe	—	—	0.64	0.95	0.84	—
Ga	—	—	—	—	—	—
Ge	—	—	—	—	—	—
Hg	—	—	(II) 0.91	—	—	—
Hf	—	—	—	—	—	—
In	—	—	—	—	—	—
K	—	—	—	—	—	—
Li	—	—	—	—	—	—
Mg	—	—	—	—	—	—
Mn	—	—	0.63	0.10	0.20	—
Mo	—	—	—	—	—	—
Na	—	—	—	—	—	—
Nb	—	—	—	—	—	—
NH$_4$	—	—	—	—	—	—
Ni	—	—	0.61	0.00	0.50	—
Pb	—	—	0.32	—	—	—
Rb	—	—	—	—	—	—
Sb	—	—	—	—	—	—
Se	—	—	—	—	—	—
Sn	—	—	—	—	—	—
Sr	—	—	—	—	—	—
Ta	—	—	—	—	—	—
Te	—	—	—	—	—	—
Ti	—	—	—			0.38
Tl	—	—	0.21			—
Th	0.51	0.00	—			0.77
U	—	0.88	0.75	—	—	—
V	—	—	—	—	—	—
W	—	—	—	—	—	—
Zn	—	—	0.60	—	—	—
Zr	0.00	—	—	—	—	0.30
Ref.	[314]	[312]	[233]	[243]	[248]	[313]

Solvent system

Element	BuOAc + AmOH + H₂O + HCl (50:10:1:20)	PhOH + PrOH + HCl (5:3:2)	PhOH + MeOH + HCl (115:45:40)	Trichloroethylene + H₂O + MeOH + AcOH (17:8:9:30)	EtCel + H₂O + HCl (7:2:1)	C_6H_6 + MeOH + AcOH (93:5:2)
Ag	—	—	—	0.48	—	—
Al	—	0.00	0.10	—	—	—
Au	1.00	—	—	—	—	—
As	0.85	—	—	—	—	—
Ba	—	—	0.00	—	0.12	—
Be	—	—	0.23	—	0.77	—
Bi	—	—	0.20	0.81	—	—
Ca	—	—	0.70	—	0.35	—
Cd	—	—	0.30	—	—	—
Co	0.50	—	0.16	—	—	0.45
Cr	0.30	—	0.12	—	—	—
Cs	—	—	0.37	—	0.61	—
Cu	0.59	—	0.22	0.68	—	0.92
Fe	1.00	—	0.39	0.75	—	0.92
Ga	—	0.80	0.66	—	—	—
Ge	—	—	—	—	—	—
Hg	—	—	(II) 0.34	(II) 0.86	—	—
Hf	—	—	—	—	—	—
In	—	0.06	0.20	—	—	—
K	—	—	0.11	—	0.61	—
Li	—	—	0.30	—	0.31	—
Mg	—	—	0.17	—	0.50	—
Mn	—	—	0.16	—	—	0.14
Mo	—	—	—	—	—	—
Na	—	—	0.11	—	0.46	—
Nb	—	—	—	—	—	—
NH₄	—	—	0.25	—	—	—
Ni	0.31	—	0.14	—	—	0.89
Pb	0.35	—	—	0.59	—	—
Rb	—	—	0.17	—	0.61	—
Sb	1.00	—	(III) 0.44	—	—	—
Se	—	—	—	—	—	—
Sn	—	—	—	—	—	—
Sr	—	—	0.03	—	0.22	—
Ta	—	—	—	—	—	—
Te	—	—	—	—	—	—
Ti	—	—	(IV) 0.37	—	—	—
Tl	—	0.67	0.53	—	—	—
Th	—	—	—	—	—	—
U	0.61	—	—	—	—	—
V	0.32	—	—	—	—	—
W	—	—	—	—	—	—
Zn	—	—	0.35	—	—	—
Zr	—	—	—	—	—	—
Ref.	[517]	[287]	[241]	[515]	[147]	[235]

Solvent system

Element	C_6H_6 + MeOH + AcOH (93:5:2)	DX + HCl + antipyrine (100:1:1)	100 ml DX + 2 ml HNO_3 + 0.05 g EDTA in 2 ml H_2O + 1 g antipyrine	$CHCl_3$ + EtOH + HCl (17:7:1)	$CHCl_3$ + MeOH + H_2O + AcOH (35:7:8:25)	COL satd. with 0.4M HNO_3	COL + H_2O (1:1)	PY + H_2O (6:4)
Ag	—	—	0.08	—	0.41	0.78	0.90	0.88
Al	—	—	0.16	0.10	—	0.00	—	0.00
Au	—	—	—	—	—	—	—	—
As	—	—	(III) 0.43	0.85	—	(III) 0.66	0.70	0.75
Ba	—	—	0.02	—	—	0.27	—	—
Be	—	—	0.26	0.30	—	—	—	—
Bi	—	—	0.69	—	0.80	0.00	—	0.00
Ca	—	—	0.17	0.10	—	0.55	—	—
Cd	—	—	0.24	—	—	0.77	—	0.87
Co	0.45	—	0.12	0.06	—	0.75	0.48	0.88
Cr	—	—	(III) (VI) 0.04 0.03	(III) 0.03	—	0.00	—	0.00
Cs	—	—	—	—	—	—	—	—
Cu	0.92	—	0.29	0.27	0.63	0.77	0.70	0.85
Fe	0.92	—	0.23	(III) 0.95	0.88	0.00	0.00	0.00
Ga	—	—	—	—	—	—	—	—
Ge	—	—	—	0.60	—	—	—	—
Hg	—	—	(II) 0.55	—	(II) 0.78	0.00	—	0.86
Hf	—	—	—	—	—	—	—	—
In	—	—	—	—	—	—	—	—
K	—	majori-	0.04	—	—	0.32	—	—
Li	—	ty of	0.28	—	—	—	—	—
Mg	—	$R_F <$	0.13	—	—	0.65	—	—
Mn	0.14	0.40	0.15	0.03	—	0.72	0.28	0.87
Mo	—	—	(VI) 0.42	0.82	—	—	—	—
Na	—	—	0.06	—	—	0.42	—	—
Nb	—	—	—	—	—	—	—	—
NH_4	—	—	—	—	—	—	—	—
Ni	0.89	—	0.12	0.04	—	0.78	0.55	0.87
Pb	—	—	0.17	—	0.54	0.00	0.10	—
Rb	—	—	—	—	—	—	—	—
Sb	—	—	(III) 0.00	(III) (V) 1.00 0.95	—	(III) 0.38	0.00	0.72
Se	—	—	—	—	—	—	—	—
Sn	—	—	(II) 0.82	—	—	0.00	—	0.00
Sr	—	—	0.06	—	—	0.40	—	—
Ta	—	—	—	—	—	—	—	—
Te	—	—	—	—	—	—	—	—
Ti	—	—	(IV) 0.11	0.07	—	—	—	—
Tl	—	—	—	—	—	—	—	—
Th	—	0.78	0.73	0.04	—	—	—	—
U	—	0.81	0.75	0.17	—	—	—	—
V	—	—	(V) 0.17	0.09	—	—	—	—
W	—	—	—	0.00	—	—	—	—
Zn	—	—	0.12	—	—	0.76	—	0.86
Zr	—	—	—	0.00	—	—	—	—
Ref.	[235]	[36]	[513]	[276]	[517]	[124]	[96]	[95]

Solvent system

Element	PY + AcOH + HCl (6 : 80 : 20)	0.2M H_2SO_4	0.4 M H_2SO_4	0.5M H_2SO_4	6M HCOOH
Ag	—	0.92	—	—	—
Al	0.30	—	—	—	1.00
Au	—	—	—	—	—
As	—	—	—	—	—
Ba	—	—	—	—	—
Be	—	—	—	—	—
Bi	—	—	0.88	—	—
Ca	—	—	—	—	—
Cd	—	—	—	0.93	--
Co	0.68	--	—	—	—
Cr	0.26	—	—	—	—
Cs	—	—	—	—	—
Cu	—	—	—	0.82	—
Fe	0.92	—	—	--	0.85
Ga	—	—	—	—	—
Ge	—	—	—	—	—
Hg	--	0.98	—	(II) 0.64	—
Hf	—	—	—	—	—
In	—	—	—	—	—
K	—	—	—	—	—
Li	—	—	—	—	—
Mg	—	—	—	—	—
Mn	0.57	—	—	—	—
Mo	—	—	—	—	—
Na	—	—	—	—	—
Nb	—	—	—	—	—
NH_4	—	—	—	—	—
Ni	0.39	—	—	—	—
Pb	—	0.17	0.00	0.00	—
Rb	—	—	—	—	—
Sb	—	—	—	—	—
Se	—	—	—	—	—
Sn	—	—	—	—	—
Sr	—	—	—	—	—
Ta	—	—	—	—	—
Te	—	—	—	—	—
Ti	—	—	—	—	0.82
Tl	—	—	—	—	—
Th	—	—	—	—	—
U	—	—	—	—	—
V	—	—	—	—	—
W	—	—	—	—	—
Zn	0.82	—	—	—	—
Zr	—	—	—	—	—
Ref.	[213]	[85]	[300]	[236]	[288]

Solvent system

Element	4M NH₃	0.6M NH₃ (complex sulphides diss. in NH₃)	7.5M LiCl + 10M-HCl (99+1)	0.2M (NH₄)₂CO₃ 0.05M NH₄HCO₃	satd. NH₄Cl dil. 1:10 + NH₄OH (s. g. 0.96) (4:1)	1.0M HCl (paper impregn. with zinc phosphate)	1.0M HNO₃ (paper impregn. with zinc phosphate)
Ag	0.28	—	—	—	—	—	—
Al	—	—	—	—	—	—	—
Au	—	—	0.27	—	—	—	—
As	—	0.75	—	—	—	(III) (V) 0.66 0.14	(III) (V) 0.25 0.57
Ba	—	—	—	—	—	—	—
Be	—	—	—	—	—	—	—
Bi	0.00	—	—	—	—	—	—
Ca	—	—	—	—	—	—	—
Cd	0.85—1.0	—	—	—	—	—	—
Co	—	—	—	—	0.85	—	—
Cr	—	—	—	—	—	—	—
Cs	—	—	—	—	—	—	—
Cu	0.85—1.0	—	—	—	—	—	—
Fe	—	—	0.88	—	—	—	—
Ga	—	—	0.55	—	—	—	—
Ge	—	—	—	—	—	—	—
Hg	0.85—1.0	—	—	—	—	—	—
Hf	—	—	—	—	—	—	—
In	—	—	—	—	—	—	—
K	—	—	—	—	—	—	—
Li	—	—	—	—	—	—	—
Mg	—	—	—		—	—	—
Mn	—	—	—	Other ions have maximum R_f 0.76	0.00	—	—
Mo	—	—	—		—	—	—
Na	—	—	—		—	—	—
Nb	—	—	—		—	—	—
NH₄	—	—	—		—	—	—
Ni	—	—	—		0.95	—	—
Pb	0.00	—	—		—	—	—
Rb	—	—	—		—	—	—
Sb	—	0.00	—		—	—	—
Se	—	—	—		—	—	—
Sn	—	0.86	—		—	—	—
Sr	—	—	—		—	—	—
Ta	—	—	—		—	—	—
Te	—	—	—		—	—	—
Ti	—	—	—		—	—	—
Tl	—	—	—		—	—	—
Th	—	—	—		—	—	—
U	—	—	—	1.00	—	—	—
V	—	—	—	—	—	—	—
W	—	—	—	—	—	—	—
Zn	—	—	—	—	—	—	—
Zr	—	—	—	—	—	—	—
Ref.	[300]	[103]	[74]	[321]	[328]	[106]	[106]

Solvent system

Element	0.1M HCl (asbestos paper washed with HCl)	0.9M NH$_4$Cl (paper impregn. with zirconium molybdate)	0.1M HClO$_4$ (paper impregn. with ammonium phosphomolybdate)	0.2M HNO$_3$ + 3.5M NH$_4$NO$_3$ (paper imptegn. with ammonium phosphomolybdate)	10M HCl (paper impregn. with tri-n-octylamine)	1M HCl (paper impregn. with di-(2-ethylhexyl)-phosphoric acid)	2M NH$_4$NO$_3$ (paper impregn. with 0.1M-tri-n-octylamine)
Ag	—	—	—	—	—	—	—
Al	—	—	—	—	—	0.00	—
Au	—	—	—	—	—	—	—
As	—	—	—	—	—	—	—
Ba	0.76	0.71	—	—	—	—	—
Be	—	—	—	—	—	—	—
Bi	—	—	—	—	—	—	—
Cs	—	0.18	—	—	—	—	—
Cd	—	—	—	—	—	—	—
Co	—	—	—	—	—	—	—
Cr	—	—	—	—	—	—	—
Ca	0.45	—	0.00	0.10	—	—	—
Cu	—	—	—	—	—	—	—
Fe	—	—	—	—	—	—	—
Ga	—	—	—	—	—	0.92	—
Ge	—	—	—	—	—	—	—
Hg	—	—	—	—	—	—	—
Hf	—	—	—	—	—	—	—
In	—	—	—	—	—	0.32	—
K	0.77	—	0.32	0.86	—	—	—
Li	0.99	—	0.80	0.90	—	—	—
Mg	—	—	—	—	—	—	—
Mn	—	—	—	—	—	—	—
Mo	—	—	—	—	—	—	—
Na	0.86	—	0.74	0.89	—	—	—
Nb	—	—	—	—	—	—	—
NH$_4^+$	—	—	—	—	—	—	—
Ni	—	—	—	—	—	—	—
Pb	—	—	—	—	—	—	—
Rb	0.72	—	0.03	0.60	—	—	—
Sb	—	—	—	—	—	—	—
Se	—	—	—	—	—	—	—
Sn	—	—	—	—	—	—	—
Sr	0.90	0.48	—	—	—	—	—
Ta	—	—	—	—	—	—	—
Te	—	—	—	—	—	—	—
Ti	—	—	—	—	—	—	—
Tl	—	—	—	—	—	0.78	—
Th	—	—	—	—	0.95	—	0.10
U	—	—	—	—	0.00	—	0.51
V	—	—	—	—	—	—	—
W	—	—	—	—	—	—	—
Zn	—	—	—	—	—	—	—
Zr	—	—	—	—	0.35	—	—
Ref.	[178]	[181]	[215]	[215]	[266]	[284]	[317]

Solvent system

Element	MeOH (paper impregn. with 4-hydroxybenzothiazole)	Dioxan (paper impregn. with oxine)	Pyridine (paper impregn. with oxine)	CHCl₃ (paper impregn. with oxine)	MeOH (paper impregn. with oxine)	0.1M HCl (paper impregn. with zirconium phosphate)	1M HCl (paper impregn. with zirconium phosphate)	Me₂CO + H₂O + AcOH (7:2:1)
Ag	—	—	—	—	—	—	—	—
Al	0.65	0.79	0.96	0.65	0.65	0.13	0.77	—
Au	—	—	—	—	—	—	—	—
As	—	—	—	—	—	—	—	(III) 0.44
Ba	—	0.68	0.00	—	0.46	0.60	—	—
Be	—	—	—	—	—	—	—	—
Bi	—	—	—	—	—	—	0.83	0.50
Ca	—	—	0.11	0.00	0.42	0.81	—	—
Cd	0.22	0.00	0.91	0.01	0.14	0.60	0.81	0.60
Co	0.25	0.94	0.87	0.70	0.86	0.67	0.77	—
Cr	—	—	—	—	—	0.75	—	—
Cs	—	—	—	—	—	0.00	0.00	—
Cu	0.25	0.81	0.91	0.62	0.65	0.59	0.78	0.30
Fe	0.59	0.99	0.92	0.74	—	(III) 0.0	0.04	—
Ga	—	—	—	—	—	—	—	—
Ge	—	—	—	—	—	—	—	—
Hg	—	—	—	—	—	(II) 0.67	—	0.96
Hf	—	—	—	—	—	—	—	—
In	—	—	—	—	—	—	—	—
K	—	—	—	—	—	0.53	0.60	—
Li	—	—	—	—	—	0.82	0.85	—
Mg	—	—	0.53	0.00	0.60	0.76	—	—
Mn	—	—	—	—	—	0.60	—	—
Mo	—	—	—	—	—	—	—	—
Na	—	—	—	—	—	0.68	0.83	—
Nb	—	—	—	—	—	—	—	—
NH₄	—	—	—	—	—	—	—	—
Ni	0.15	0.79	0.93	0.71	0.79	0.61	0.85	—
Pb	—	0.69	0.83	0.00	—	—	—	0.18
Rb	—	—	—	—	—	0.12	0.48	—
Sb	—	0.00	0.90	—	0,69	—	—	C
Se	—	—	—	—	—	—	—	—
Sn	—	—	—	—	—	—	—	—
Sr	—	—	—	—	—	0.81	—	—
Ta	—	—	—	—	—	—	—	—
Te	—	—	—	—	—	—	—	0.12
Ti	—	—	—	—	—	0.00	—	—
Tl	—	—	—	—	—	0.00	0.45	—
Th	—	—	—	—	—	0.00	0.00	—
U	—	—	—	—	—	0.00	0.32	—
V	—	—	—	—	—	—	—	—
W	—	—	—	—	—	—	—	—
Zn	0.10	—	—	—	—	0.69	0.76	—
Zr	—	—	—	—	—	—	—	—
Ref.	[360]	[514]	[514]	[514]	[514]	[518]	[519]	[90]

EXPLANATION OF THE TRIVIAL NAMES
AND ABBREVIATIONS USED FOR COMPOUNDS

1	Alizarin	1,2-dihydroxyanthraquinone
2	Alizarin Red S	sodium 1,2-dihydroxyanthraquinone-3-sulphonate
3	Aluminon	ammonium salt of aurinetricarboxylic acid
4	Anthranilic acid	*o*-aminobenzoic acid
5	Antipyrine	2,3-dimethyl-1-phenylpyrazol-5-one
6	Arsenazo	sodium salt of *o*-(1,8-dihydroxy-3,6-disulpho-2-naphthylazo)benzenarsonic acid
7	Bismuthiol II	potassium salt of 2-mercapto-3-phenyl-2-thiol-1,3,4-thiadiazol-2-one
8	Bromocresol Red	dibromo-*o*-cresolsulphonphthalein
9	Bromocresol Green	tetrabromo-*m*-cresolsulphonphthalein
10	Bromothymol Blue	dibromothymolsulphonphthalein
11	Catechol Violet	3,3′,4′-trihydroxyfuchsone-2″-sulphonic acid
12	Chromazurol S	sodium salt of 3-sulpho-2,6-dichloro-3,3′-dimethyl-4-hydroxyfuchsone-5,5′-dicarboxylic acid
13	Chromotropic acid	1,8-dihydroxynaphthalene-3,6-disulphonic acid
14	Cupferron	*N*-nitrosophenylhydroxylamine, ammonium salt
15	Dithiol	4-methyl-1,2-dimercaptobenzene
16	Dithizone	diphenylthiocarbazone
17	EDTA	ethylenediaminetetra-acetic acid (usually as the disodium salt)
18	Eriochromecyanine R	di- or trisodium salt of 5-[α(3-carboxy-5-methyl-4-oxo-2,5-cyclohexadien)1-ylidene)-2-sulphobenzyl]-3-methylsalicylic acid
19	Gallion	2-hydroxy-5-nitrobenzene-(1-azo-2-)-1′-hydroxy-8′-alizarinophthalene-3′,6′-disulphonic acid
20	Kojic acid	2-hydroxymethyl-5-hydroxy-γ-pyrone
21	Magneson I	*p*-nitrobenzeneazo-resorcinol
22	Magneson II	*p*-nitrobenzeneazo-1-naphtol
23	Malachite Green	tetramethyl-di-*p*-diaminofuchsonium chloride
24	Mercupral	copper salt of tetraethylthiuram disulphide
25	Morin	3,5,7,2′,4′-pentahydroxyflavone
26	Nitroso-R-salt	sodium 1-nitroso-2-hydroxynaphthalene-3,6-disulphonate
27	Phenol Red	phenolsulphonphthalein
28	Quercetin	3,5,7,3′,4′-pentahydroxyflavone
29	Quinalizarin	1,2,5,8-tetrahydroxyanthraquinone
30	Rubeanic acid	dithio-oxamide
31	Tiron	disodium 4,5-dihydroxy-*m*-benzenedisulphonate
32	Titan Yellow	sodium methylbenzothiazole-1,3,4,4′-diaminobenzene-2,2′-disulphonate
33	Thymol Blue	thymolsulphonphthalein
34	Violuric acid	5-isonitrosobarbituric acid
35	Zincon	2-carboxy-2′-hydroxy-5-sulphoformazylbenzene

1	AcMe$_2$CO	acetylacetone
2	AcOH	acetic acid
3	AllOH	allyl alcohol
4	AmOH	1-pentanol
5	BuOAc	butyl acetate
6	BuOH	1-butanol
7	COL	collidine
8	DX	dioxan
9	EtCel	monoethyl cellosolve (ethylene glycol monoethyl ether)
10	Et$_2$CO	diethyl ketone
11	Et$_2$O	diethyl ether
12	EtOAc	ethyl acetate
13	EtOH	ethanol
14	Hcit	citric acid
15	HCOOH	formic acid
16	Hox	oxalic acid
17	Htar	tartaric acid
18	iBuOH	isobutyl alcohol
19	iPrOH	2-propanol
20	Me$_2$CO	acetone
21	MeEtCO	ethyl methyl ketone
22	MeiBuCO	methyl isobutyl ketone
23	MenPrCO	methyl n-propyl ketone
24	MeOAc	methyl acetate
25	MeOH	methanol
26	nPrOH	1-propanol
27	PhCH$_2$Ac	benzylacetone
28	PhCH$_2$OH	benzyl alcohol
29	PhOH	phenol
30	PY	pyridine
31	ROAc	acetic acid ester
32	ROAcAc	acetoacetic acid ester
33	tBuOH	t-butyl alcohol
34	THF	tetrahydrofuran
35	THS	tetrahydrosylvan

INDEX